D0056663

AROUSED

ALSO BY RANDI HUTTER EPSTEIN

*Get Me Out: A History of Childbirth from the
Garden of Eden to the Sperm Bank*

AROUSED

The History of Hormones and How They
Control Just About Everything

Randi Hutter Epstein, MD, MPH

W. W. NORTON & COMPANY

Independent Publishers Since 1923

New York | London

For information about permission to reproduce selections from this book,
write to Permissions, W. W. Norton & Company, Inc.,
500 Fifth Avenue, New York, NY 10110

For information about special discounts for bulk purchases, please contact
W. W. Norton Special Sales at specialsales@wwnorton.com or 800-233-4830

Manufacturing by LSC Communications, Harrisonburg
Book design by Ellen Cipriano
Production manager: Lauren Abbate

Library of Congress Cataloging-in-Publication Data

Names: Epstein, Randi Hutter, author.
Title: Aroused : the history of hormones and how they control
just about everything / Randi Hutter Epstein.
Other titles: History of hormones and how they control just about everything
Description: First edition. I New York : W. W. Norton & Company, 2018 I
Includes bibliographical references and index.
Identifiers: LCCN 2017061090 I ISBN 9780393239607 (hardcover)
Subjects: I MESH: Hormones—history I Hormones—physiology
Classification: LCC QP571 I NLM WK 11.1 I DDC 612.4/05—dc23
LC record available at https://lccn.loc.gov/2017061090/

W. W. Norton & Company, Inc., 500 Fifth Avenue, New York, N.Y. 10110
www.wwnorton.com
W. W. Norton & Company Ltd., 15 Carlisle Street, London W1D 3BS

1 2 3 4 5 6 7 8 9 0

For Stuart
&
Jack, Martha, Joey, Eliza

Contents

INTRODUCTION   xi

1 The Fat Bride   1
2 Hormones . . . As We May Call Them   14
3 Pickled Brains   34
4 Killer Hormones   54
5 The Virile Vasectomy   72
6 Soul Mates in Sex Hormones   89
7 Making Gender   102
8 Growing Up   126
9 Measuring the Immeasurable   148
10 Growing Pains   160
11 Hotheads: The Mysteries of Menopause   174
12 Testosterone Endopreneurs   194
13 Oxytocin: That Lovin' Feeling   214
14 Transitioning   230
15 Insatiable: The Hypothalamus and Obesity   246
   Epilogue   257

ACKNOWLEDGMENTS   261
NOTES   267
INDEX   299

Introduction

IN THE SUMMER OF 1968, I spent a lot of time at my grandmother's pool, the Sprain Brook Country Club in Yonkers, New York. My grandmother Martha and her three friends (always the same foursome) sat in the shade, playing canasta, drinking hot coffee, and smoking Kents.

I'd swim with my older brother and sister, but mostly my sister and I sunbathed, our bodies slathered in Johnson's baby oil, our heads tucked into the crack of an album cover that we had wrapped in aluminum foil, trying to grab the sun's rays.

On the way home, my sister and I would hold our arms side by side. She always had a decent tan; I, a redhead, was consistently the color of a ripe tomato, the kind of burn that peels off the next day. But Grandma Martha was spectacularly bronzed. She seemed to absorb the best rays without even trying.

Five years later, we learned that our grandmother didn't have a particular aptitude for sunbathing. She had a hormone problem: Addison's disease. Her body wasn't producing enough corti-sol, a hormone that helps maintain a healthy blood pressure and strengthens the immune system. People with Addison's disease suffer from extreme fatigue, nausea, and low blood pressure, some-times dangerously low. The disease also darkens skin. Once she was diagnosed, the treatment was easy. She took daily cortisone pills, which contain a hormone chemically similar to cortisol, the one she was missing.

When my grandmother was born in 1900, the word "hormone" didn't exist. It was coined in 1905. By the time she got sick, in the 1970s, scientists had a way to spot her hormone defect, measure her hormones down to a billionth of a gram, and prescribe pills that kept her illness at bay.

In 1855, Claude Bernard, a renowned physiologist, had a hunch that the liver had something to do with preventing wild swings in sugar levels in the body. He had been studying digestion and had already discovered that the pancreas releases juices that break down food. To test this, he fed a dog a diet of meat and no sugar. Then he killed it, immediately removed its liver and tested the still-warm organ for sugar, a few minutes later and again several hours after that. Much to his delight, the sugar level in the dog's liver started at practically zero but continued to rise. (Even though the dog was dead, the liver—as do other organs—kept functioning for a few days; that's why transplants work.)

Bernard announced to his colleagues that the liver must contain a chemical that stores and produces sugar. But he also proclaimed that all organs, not just the liver and pancreas, release substances that keep the body running smoothly. He called these chemicals "internal secretions." It was a new way of thinking about the body.

Many historians consider Bernard the father of endocrinology.

I don't. The real pioneers recognized that these chemicals aren't just internal secretions. They play a more important role. They arouse. They excite receptors on target cells, flipping switches that get things going.

I delved into the history of hormones because the last century has been a period of incredible discovery and also outrageous claims. In the 1920s, the discovery of insulin and its use as a treatment changed diabetes from a death sentence to a chronic disease. In the 1970s, a test for thyroid hormone among newborns prevented thousands of children from growing up intellectually disabled. At the same time, there were wild missteps. Vasectomies were promoted as a treatment to rejuvenate elderly men, a fad that lasted nearly a decade beginning in the mid-1920s. Not long after that, a doctor who treated the literati claimed he could detect hormone ailments by studying people's faces, and prescribed hormone-based remedies. It was hocus-pocus combined with potent and potentially dangerous treatments.

Aroused tells the stories of daring scientists and also of desperate parents. In the days before sophisticated imaging techniques, an early twentieth-century neurosurgeon performed brain operations to remove bits of a gland that he thought would staunch diseases caused by an overdose of hormones. In the 1960s, one couple scoured pathology labs and morgues to get growth hormone for their short son. *Aroused* is also the story of curious shoppers, dying (sometimes literally) to tap into the hormone hype in order to live a little longer or feel a little better. I begin with doctors in the late 1800s scrounging the glands of cadavers, some of them stolen from graveyards. I culminate with scientists tracking hormone pathways right down to the genes that make them.

How did we discover that growth hormone isn't just for growing? When did we learn that the testes and ovaries are controlled by a hormone in the brain? Does the recently discovered hunger hormone mean it's not really my lack of willpower but my own chemistry nudging me to binge? And if so, is there really a difference between the two? I am my chemistry, after all. And what do the newest studies say about hormones being used today: the testosterone gels popular among aging men and hormone replacement therapy for menopausal women?

Aroused starts in the prequel to hormones, when nineteenth-century medical practitioners began to eye the chemical-secreting glands scattered throughout the body. That research led to the concept of hormones in the early 1900s. By the 1920s, the field—endocrinology—had exploded from an obscure science to one of the most widely discussed medical specialties. In addition to the discovery of insulin, estrogen and progesterone were isolated. At the same time, advice books flourished promoting all kinds of wacky remedies.

If the Roaring Twenties was endocrinology's coming-out party, the era when it gained popularity for both real and quack cures, the 1930s cemented its role as a serious science. Three pivotal advances in biochemistry debunked dogma of years past. It had been thought that estrogen and testosterone were wildly different substances, but researchers in this decade discovered that they differ by only one hydroxyl group: that's just one hydrogen atom and one oxygen atom. Estrogen and testosterone are basically fraternal twins in different clothing. Secondly, when estrogen was finally isolated, from horse urine, it came not from females but from excretions of stallions. Scientists had assumed that ovaries make estrogen and testicles make testosterone; this discovery prompted scientists to realize that both make both. And lastly, investigators had thought that estrogen and testosterone were antagonistic chemicals: like kids on a seesaw, the rise of one pushed the other down. But really, the two chemicals are not antagonists at all, but partners that often work in cahoots.

These findings ushered in a more complicated view of hormones. Scientists were no longer studying one at a time, but examining how they were connected.

The second half of the twentieth century began with triumph. Scientists figured out how to measure hormones, something that had been considered impossible. That's because despite their power, hormones come in tiny packets. They had been considered too scarce to measure. Shortly thereafter, the birth control pill was approved; a take-home speedy pregnancy test replaced older, slower methods; and bottled hormones were sold to quash the symptoms of menopause. But the glee didn't last long. As hormone drugs became hugely popular, side effects began to emerge. The original dose of the pill was linked to deadly strokes. Hormone replacement therapy, once assumed to prevent all kinds of chronic ailments of old age, was found to be effective but not the wonder cure it had been touted to be. Today we are taking a more discerning approach to hormone therapies, but much remains to be known.

How do we weigh benefits with potential risks? The point is not to promote a new way to stay young forever (that's an old story and one that continues to be rewritten), nor is it to promote all things natural (we are made of hormones, after all; they are our natural chemistry). Rather, *Aroused* helps readers appreciate the complex interactions that hormones have with each other within our bodies and the relationships we have with hormones to which we are exposed.

It was only recently that my mother told me that in the weeks leading up to Grandma Martha's diagnosis, the card-playing ladies said that my grandmother was unusually exhausted. She'd fall asleep during games. Then, on the Monday before Thanksgiving in 1974, she showed up at our house in New Jersey and sat there calmly. Instead of dipping a spoon into the soup, crinkling her nose and muttering under her breath that it needed more salt, she rested on a couch. This was not the Grandma Martha we knew. (Salt craving, we would soon learn, was another sign of Addison's disease.) There was no Grandma Martha gossiping, no complaining. She didn't even have the energy to go out onto the back porch and smoke a cigarette. My mother got scared and called the doctor.

He couldn't find anything wrong, but given the odd personality change, he admitted my grandmother to the hospital for extra tests. By the time my grandmother was wheeled to her assigned bed, she was too weak to lift a fork, so my mother had to feed her. That's when my mother noticed that my grandmother's tongue was pitch black. (In retrospect, my mother isn't sure if the internist ever examined her. How did he miss the symptoms?)

My father, a pathologist, put the clues together—black tongue, bronzed skin, extreme fatigue—and suspected Addison's disease. He insisted on hormone testing. It revealed a dearth of cortisol.

Back then I didn't know much about her disease except that John F. Kennedy had the same thing, which made it sound like a very presidential thing to catch. My memory throughout my childhood is my mother saying, "Ma, don't forget to take your cortisone pill." There was one pill in the morning and one in the afternoon. I'm not even sure I knew Addison's was a hormone disease. To me, hormones were boobs and periods and sex. Simple enough.

But hormones are so much more. They are the potent chemicals that control metabolism, behavior, sleep, mood swings, the immune system, fighting, fleeing—not just puberty and sex. So in a sense this is a story of the biochemistry of living, breathing, emotional beings. The history of hormones is also a story of discovery, wrong turns, persistence, and hope. Taking both together—the basic science and the people who shaped it—*Aroused* is a tale of what makes us human, from the inside out.

AROUSED

1.

The Fat Bride

NOT EVEN A DAY had passed since the Fat Bride was dead and buried before the body snatchers tried to dig her up and cart her off to scientists. The first attempted exhumation occurred around midnight on October 27, 1883, at Baltimore's Mount Olivet Cemetery. The cemetery's nightwatchman fired his gun, startling the vandals, who fled with their spades and shovels. An hour later, gunshots sent another group running from the same grave. Newspapers offered conflicting stories. Some said bullets pierced two grave robbers. Others said no one was hurt. In any event, everyone survived. Except, of course, the bride.

It's a wonder anyone thought they could exhume all 517 pounds of Blanche Grey, a.k.a. the Fat Bride. For starters, it had taken a dozen burly men to strap her body to a wooden plank, carry her down three flights of stairs, lift her onto the undertaker's wagon, and plant her six feet under. It would require at least that many men to hoist her up. Secondly, her body was a coveted medical commod-

Resurrectionists, from the Healy Collection, New York Academy of Medicine. *Courtesy of the New York Academy of Medicine Library.*

ity, so the guard was particularly vigilant that night, maintaining a lookout from the second-floor window of his house on the cemetery grounds, within viewing distance of the plot. A coworker helped; together they took turns staring out the window with their guns in ready position.

Poor Blanche Grey. She was born in Detroit, a huge baby, a twelve-pounder, who ballooned to 250 pounds by the time she turned twelve. Her mother had died a few days after her birth; her father and two brothers figured no one would marry her and she'd be stuck at home forever. Grey thought otherwise. She was determined to get as far away as possible, to start a new life for herself that was under neither the critical gawk of her family nor the curious gaze of doctors. She chose a profession that put her in the spotlight anyhow.

At seventeen, Grey boarded a bus and headed to Manhattan to get a job in a freak show. She reckoned she could get a Fat Lady gig, sitting alongside the other "abnormals"—the bearded ladies, the midgets, the giants, and the rest of them. Sometimes they arranged themselves in a cavernous room; sometimes they were shoved behind an amusement park's roller coaster. Savvy entrepreneurs promoted these voyeuristic shows as educational entertainment.

The display of so much human variation crammed into one place didn't just amuse a prurient public; it stoked the curiosity of an eclectic crew of physiologists, neuroscientists, and biochemists. They aimed to prove that these folks were the way they were because of a physical defect, not because of a moral failing or divine punishment as was generally thought. If they could figure out what made these folks dangerously different, they just might discover what made the rest of us so wonderfully normal.

If Grey had been born a hundred years later, living in the second half of the twentieth century instead of the nineteenth century, doctors might have tested her for various hormone defects linked to obesity—problems, perhaps, with the levels of thyroid and growth hormones. If she had been born closer to 2000, there's a good chance she would have seen endocrinologists who might have analyzed her levels of leptin and grehlin. The doctors who examined her at birth might have suspected that her mother suffered from diabetes, a hormone disease that, among other things, increases the odds of having an obese baby. They would have known enough about hormone problems to check for other conditions, too. Untreated thyroid deficiencies at birth, for instance, lead not only to weight gain but also to cognitive deficiencies and dry skin.

But Grey lived on the wrong side of scientific discovery.

There were hints. Forty years before Grey's death, in 1840, an autopsy of a woman dead from so-called fatal obesity revealed a tumor encroaching on a brain gland. Not long after that, the corpse of an obese, developmentally delayed ten-year-old was found to lack a throat gland. Had a gland disease killed Grey?

Almost as soon as Grey arrived in New York City, she was earning $25 a week as a Fat Lady in a museum at 210 Bowery. (The museum would become the Monroe Hotel for Skid Row bums in the 1930s and then a luxury high-rise in 2012.) Just as quickly she caught the eye of David Moses, the ticket taker, who pulled in a relatively meager $5 weekly salary. After a few dates, he proposed to become her husband and manager. She said yes to both. She was seventeen, but said she was eighteen. He was twenty-five. Moses sold tickets to their wedding at New York City's Dime Museum. A huge banner flapped across the entrance: "Blanche Grey, The Fattest Girl in the World will be Married on Stage Tonight at 9pm!" Moses placed articles in local newspapers to ensure a sold-out audience. The advertisements hailed Blanche as the "wonder of the 19th century."

"A Ponderous Bride," said the *Baltimore Sun*. "More Than a Better Half," said the *New York Times*. Blanche Grey "tipped the scales at 517 pounds, so it was only natural, therefore, and in accordance with the laws of gravitation, that the smaller body should be attracted by the larger." The *Times* called her an "adipose monstrosity."

Right after the ceremony, Moses had another proposition: to switch Grey's stage name from Fat Lady to Fat Bride. He said it would give her an advantage in an increasingly crowded field. Fat Girls and Ladies, after all, were generic. Fat Brides were rare. Moses guaranteed exhibitors swarms of paying gawkers because Blanche's betrothal and wedding had been such a media sensation. He scored a slew of bookings for their working honeymoon. The morning after the reception, the new Mrs. Moses had a "performance" at New York City's bustling Coney Island boardwalk. From there, her husband booked Baltimore's Dime Museum and then Philadelphia's Hagar & Campbell Casino.

At first, everything seemed to be going stupendously well. The Dime Museum in Baltimore threw in free rooms at a boarding house not just for the newlyweds but for their wedding attendants as well: the Armless Dwarf, the Bearded Lady, and the White Moor. (The locals called it the "freaks' boarding house.") The only almost-glitch was their honeymoon suite: Blanche had trouble managing the ascent to the third floor. The museum agreed to pulley her up—with the help of men and a crane. Moses suggested they sell tickets for her vertical voyage.

Within days there were ominous signs. Audiences complained that the Fat Bride had trouble keeping her eyes open. The Bearded Lady worried because Blanche's skin looked blotchy and purple. Her husband would later say that he kept an eye on her but never realized how sick she was. Despite outward appearances, the *Baltimore Sun* reported that Blanche was "jolly and happy," and even "encouraged the eye winking of the living skeleton at the museum much to the dislike of her husband who grew jealous."

A few days later, she died. Moses was stunned. He had slept through the night, waking up around seven o'clock because his wife rolled over. She was breathing heavily, so he kissed her and went back to sleep. An hour later he was startled by a knock at the door.

It was the manager. Before he got up, he said, he stopped to look at his wife and realized she was dead.

Her death made headlines, as her wedding had done: "The Fattest of Brides Dead" reported the *Baltimore Sun*. "Her Fat Killed Her" headlined the *Chicago Daily Tribune*. Her death was even covered in the *Irish Times:* "Sudden Death of a Fat Woman."

Crowds gathered as Grey was hauled to the cemetery. Women returning home from shopping dropped their baskets and stared. Young girls pushed through the throng for front-row viewing. Boys climbed telegraph poles. Neighbors leaned out of windows. As they watched the obese woman being carted from the "freaks' hotel," they got a free peek at the tearful One-Armed Woman, the Bearded Lady, and the other circus folk who walked alongside her. "The crowd on the pavement seemed to regard the grief of the poor dead woman's friends as a show to be stared at," the *Baltimore Sun* reported. "They laughed and nudged each other."

Grey's tragic tale epitomizes Gilded Age America: freak shows, a disdain for the abnormal (yet capitalizing on them), and a sensationalist press. Reportedly Moses tried to profit from Grey's death, selling photos of her dead body for a dime apiece. Despite a spate of articles that seem more parable than journalism, no one bothered to say that Grey's death was just like her life: a media brouhaha that took little account of her as a person. She was, it seemed, a gimmick for the press and a meddlesome public.

But Blanche's story is more than a tale of fleeting fame and minimal fortune; it casts a spotlight on late nineteenth-century medicine. Blanche died at the very moment scientists were beginning to unravel the mysteries of our endocrine system, the secretions from internal organs: our hormones. Why were some folks too fat? Too hairy? Too big? Too small? The discovery of hormones, just a few years after her burial, would lead to answers. And in time, understanding hormones would lead to lifesaving treatments, such as insulin for diabetes.

The research would also help us decode the chemical basis of so much about what makes us us. Not just our physical development, but our psychic development too. What triggers anger? What prompts maternal bonding? Can our internal chemistry explain hate or love or lust? Perhaps no other field of medicine tackles so vast a territory as endocrinology, the study of hormones.

Chemically speaking, hormones are loopy chains of amino acids or rings of carbon atoms with dangly bits hanging off the sides. But thinking about them in terms of design only is like describing football as an elliptical leather mass being tossed around a 100-yard-long rectangle. It doesn't express how one relatively small mass ignites such awesome power and complexity.

If you consider your body a vast information highway—a collection of messages being sent this way and that—your nervous system works like an old-fashioned operator's switchboard. It contains wires that need to be plugged into the source and the target in order to transmit signals. You can follow the path of a nerve from one end to the other. Hormones are a different story altogether. The remarkable thing about them—as opposed to all the other substances in your body—is the seemingly magical way they work. Hormones are pinballed from one cell in one part of the body and reach faraway targets—no connections needed. They are your wireless network. A brain cell, for instance, emits a hormone, just one droplet, and ignites a response in the testes or ovaries. (Other chemicals journey far, such as oxygen, which also travels in the blood. But oxygen isn't released from a gland, headed to a specific target, as hormones are.)

The nine key glands in the body, from head to genitals, are the hypothalamus, pineal, and pituitary in the brain; the thyroid and neighboring parathyroids in the throat; the islets of Langerhans in the pancreas; the adrenals, capping the kidneys; and the ovaries and testes. Scientists would discover in the early 1900s that when they removed a hormone-making gland from a dog's brain, they could inject its juices anywhere else in the body and everything would return to normal. Astonishing, really. Scientists would also learn that every one of our cells has markers, like routers for computers, that direct hormone signals precisely where they need to go.

They would also realize that hormones rarely work alone. A dip in the amount of one hormone interferes with other hormones and, like falling dominoes, throws a host of bodily functions off-kilter. All of these emissions from hormone-making glands are in some aspects different but the same. Related. Like siblings. Or maybe more like cousins.

The job of the gland is simple: to secrete hormones. The job of the hormone is trickier: to keep the body balanced.

Hormones control growth, metabolism, behavior, sleep, lactation, stress, mood swings, sleep–wake cycles, the immune system, mating, fighting, fleeing, puberty, parenting, and sex. They aim to get us back to normal when things are out of whack. And they can be the cause of commotion, too.

Endocrinology did not emerge until the nineteenth century, long after other substantial medical discoveries. By the end of the seventeenth century, scientists had recognized that the blood circulates rather than swishes back and forth, and they had a pretty good picture of human anatomy. The discovery of hormones was delayed until the birth of physiology and chemistry in the mid-1800s, which created a new way to study the body. No longer did investigators approach the body simply by exploring the terrain, like mapmakers venturing onto new land. Nor did they limit their focus to examining the routes of blood and nerves. They began to tinker with the body's chemical substances and theorize about their impact on health and disease. Medicine became more scientific. In 1894, William Osler, considered the father of modern medicine, declared that "The physician without physiology and chemistry flounders along in an aimless fashion never able to gain any accurate conception of disease, practising a sort of popgun pharmacy."

During the latter years of the twentieth century, we would learn that hormones depend on immune cells and also on chemical messengers from the brain—and vice versa. Our defense cells and brain cell messengers rely on hormones to function properly. This complex system would turn out to be more complicated than ever imagined. And we still do not fully understand it.

Back in Blanche Grey's time, researchers began to clear the fog. Medicine was in its adolescence then—bold, arrogant, naïve. Free of ethics boards and informed consent and all the other things that would reshape medical research in the late twentieth century, daring scientific sleuths thrived in a community of exploration, discovering on their own terms, with their own ideas about where to go and what to do. Their most audacious trials allowed science to progress faster than it would today, when care is taken to ensure that patients' rights are not violated.

Still, whether experiments speed or stumble along, new ideas rarely burst forth. They smolder, sometimes for decades. The theory of evolution was discussed for years before Charles Darwin publicized it in 1859. The germ theory of disease knocked about in laboratories in Europe before Robert Koch gathered definitive proof and publicized it in the 1880s. The same can be said for the discovery of hormones. (It's probably no surprise that the hormone theory emerged at the same time as the germ theory; they may be very different specialties, but they both focus on tiny things with a mighty kick.)

For centuries, healers had noted the powers of ovarian and testicular juices. They wondered about the thyroid gland in the neck and the adrenal glands that rest on the kidneys. Surely they must serve some purpose. But what?

The first genuinely scientific hormone experiment was conducted on August 2, 1848. Arnold Berthold, a doctor, performed an experiment on six roosters in his backyard in Göttingen, Germany. Many scientists then were curious about the testicles: whether they contained vital juices and how they worked. Could a testicle do its job if it were placed somewhere else in the body? Berthold cut a single testicle off of two of the roosters. He cut both testicles off another two. In the remaining two, he did an odd testicle-swapping thing, removing both testicles and re-inserting one in the belly of the other rooster. Each ended up with another bird's testicle, in the wrong place.

Here's what Berthold found: the testicle-free birds got fat and lazy and cowardly. He said they acted like hens. Their brilliant red combs faded and shrank. They stopped chasing female poultry. His uni-testicled roosters were the males, or rather the cocks, they had always been. They waddled, puffed their chests, and lusted after the hens. On autopsy, he found the lone testicle swollen. He reckoned it had swelled to compensate for the missing one.

But the most stunning finding of all, the discovery that should have shocked the testicle-research world, was the result of the gonad-switching. Berthold had wondered whether testicles could function from anywhere in the body. They did. He had implanted a testicle between the intestinal loops of a fat, lazy, castrated bird—

the young rooster, only three months old, had nothing between his drumsticks but a lone testicle in his gut—yet he turned back into a full-fledged hen-chaser, red comb and all. Berthold repeated his testicle-to-belly switch with another bird and the same thing happened. "They crowed lustily, often engaged in battle with each other and with other cockerels, and showed the usual reaction to hens," he wrote.

Berthold had assumed that when he carved his poultry, he would find a network of nerves connecting the displaced testicles to the body. Instead, he found the testicles were surrounded by blood vessels. In his four-page scientific report, Berthold explained for the first time how hormones work, writing that his experiment showed that the testes released a substance into the blood that was carried off to the rest of the body and reached a specific destination. He was right: hormones are released in one part of the body and reach a specific target like a well-aimed arrow from a bow. (He didn't use the word "hormone," because it wouldn't be coined for another half century.) No one listened. The specialty of hormonal science could have started right then and there. But it didn't.

Science is not just about doing the experiment. It is also about pursuing leads. Seeing the clues. Understanding the significance. Hammering away at hunches. Berthold's backyard rooster trial could have been *the* paradigm-changing experiment, transforming the way scientists looked at internal secretions. He published his insights in *Mueller's Archives of Anatomy and Physiology*, under the title "Transplantation der Hoden" (*hoden* is German for testicles). Then, without fanfare, he moved on to other projects. As Albert Q. Maisel wrote in *The Hormone Quest*, it was as if Columbus had discovered America and then went home to spend the rest of his life studying the streets of Madrid.

After Berthold, there were others who planted seeds that would one day flourish into a field called endocrinology: Thomas Blizard Curling, a surgeon in London, autopsied two obese, intellectually impaired girls (one died at the age of six, the other at ten) to see if he could find any kind of physical defect within. He found that neither had a thyroid gland, prompting him to publish an article linking a defective thyroid to mental disability. Thomas Addison, another Londoner, connected faulty adrenal glands to a syndrome that included peculiar brown spots and fatigue. It would, in time, be named for him: Addison's disease. George Oliver, a physician in northern England, fed his son sheep and cow adrenal glands that he picked up at the butcher, just to see what happened. The boy's blood pressure skyrocketed. Gleeful at his discovery, he teamed up with a London scientist and conducted dog studies that confirmed the human finding. The mysterious adrenal gland juice would be named "adrenaline."

Despite these varied experiments, no one in the nineteenth century saw the whole picture; they didn't realize that these different chemical-secreting glands shared similar features. So there wasn't a field, just a hodgepodge of scientists working away at individual glands. The adrenal people weren't talking to the testicle folks who weren't talking to the thyroid men.

It would take keen insight and partnership to bring these diverse studies into one category, figure out a common mode of action, and christen them with one name. It would take more research into men and women such as Blanche Grey, many of whom would be dug up and brought to scientific laboratories in Baltimore, New York, Boston, and London. Physiologists, neuroscientists, and chemists needed subjects, dead and alive, in order to examine glands and study the juices they excreted. And they needed to become a unified field of study: a group of scientists and physicians sharing ideas and findings, testing treatments that would bring help and sometimes cures to people in need. That would happen at the dawn of the twentieth century.

As for Blanche, she stayed six feet under, never to be exhumed to a Baltimore lab—despite several attempts by body snatchers. Had they gotten her, here's what they would have found: golden globules of fat, like piles of plump yellow fall leaves, draping her organs. A curious investigator might have peeled them away and pried the pituitary gland out of her brain or the thyroid gland from her neck. They might have noted if the gland seemed too big or too small. She probably would have become a scientific curiosity alongside the skeleton of an unusually tall person, providing fodder for later studies but not providing many answers.

2.

Hormones . . . As We May Call Them

O N NOVEMBER 20, 1907, a group of British medical students headed to Battersea to smash a statue of a dog. It was a particularly foggy night, even by London standards, so they thought they could get away with it.

The monument, seven and a half feet high was also a water fountain, with a high spout for people and a low one for pets. A bronzed brown terrier perched on a tall granite base. It was the inscription on the plinth that irked the students:

In Memory of the Brown Terrier Dog Done to Death in the Laboratories of University College in February 1903 after having endured Vivisection extending over more than Two Months and having been handed over from one Vivisector to Another Till Death came to his release. Also in memory of the 232 dogs Vivisected at the same place during the year 1902. Men and Women of England, how long shall these things be?

Turn-of-the-century animal rights activists had erected the statue, titled *The Brown Dog*, to symbolize their ire at animal experimentation. What raised the medical students' hackles was that, while the statue did not name names, they knew it was an attack on two doctors, their professors at University College London. William Bayliss and Ernest Starling had experimented on a brown terrier.

Hundreds of classmates were supposed to show up for the demolition, but at the last minute most balked. Seven young men set out from the university, in central London, across the Thames to Battersea, a working-class neighborhood. "A place to avoid if you could possibly help it," noted one historian.

They reached the south side of London and skulked toward the statue; but the closer they got, the more they feared the mission, worrying that neighbors or the police would come after them. So when they got to *The Brown Dog*, they hid behind benches and shrubs. Adolf MacGillicuddy, one of the students, leapt from the shrubbery, looked around to make sure no outsiders were watching, grabbed a crowbar, reached up as high as he could, and swung at the dog's paw. As soon as he had a toehold, MacGillicuddy heard footsteps. Police! He raced out of the park.

That's when a second batch of twenty-five medical students, the ones who had dithered, arrived in Battersea. Right place; wrong time. The first group had sneaked in as quietly as could be. The new crew was rambunctious. They might as well have announced their arrival with a megaphone. Duncan Jones, one of the second gang, swung a hammer at the dog, and as soon as he went to whack it again, two plainclothes officers nabbed him. Nine students trailed behind him to the police station, hoping to pay the fine. The police threw them all into cells.

The university paid their bail and, the next morning, the boys

Original *Brown Dog* statue in Latchmere Garden Estates. *Courtesy of the Wellcome Library, London.*

pleaded guilty to maliciously damaging a public monument, but not before putting up their defense: they were protecting the reputation of their esteemed University College London. The intent of the statue's inscription was clear: to portray researchers as animal torturers. As David Grimm put it in his book *Citizen Canine*, "centuries of angst over the souls of cats and dogs had come to a head."

Even those who supported the experiments on dogs did not condone the students' vandalism of public property. "There cannot be one standard of conduct for the ordinary individual and another for those whose parents are rich enough to pay the fees of the medical school," a local newspaper wrote. The boys were fined five pounds each and threatened with two months in prison and hard labor should they go after *The Brown Dog* again. The monument remained intact: tall and capped by a smug, proud canine.

The fiasco didn't quash the students' crusade. It increased their fury. That evening, a horde of young men swarmed into Trafalgar Square chanting, "Down with the Brown Dog!" They marched through central London brandishing canine effigies. This time around, there was no problem enlisting classmates. Throngs poured in from other medical schools, including Charing Cross Hospital, Guy's Hospital, King's College London, and Middlesex Hospital.

An elderly man who happened to be walking through central London said he felt something brush his shoulder, turned around, and realized that he had been tickled by a fluffy dog toy on a stick. Then he saw an angry mob carrying stuffed animals. What was going on?

"It's only them brown doggers, sir," a policeman said. "They's riled because their professor did something to a dawg wot's called vivispection, and the ladies they stuck up a monument to the dawg in Battersea, and they says it was tortured and that the professor he broke the law, but them young gentlemen says it's a shame and now the fat's in the fire, sir."

The hullabaloo could have easily slipped by as just another socialist uprising pitting the people against the Establishment. But historians have since realized that the so-called Brown Dog Affair had a greater impact on science than anyone realized at the time.

At the dawn of the twentieth century, William Bayliss and Ernest Starling demonstrated what no one before had appreciated: that glands—clusters of cells scattered around the body—all worked by the same mechanism. The pancreas, adrenals, thyroid, ovaries, testes, and pituitary should not, they said, be treated as different entities. Rather, they are parts of one large system. To test their ideas, Bayliss and Starling did what many scientists did then: they experimented on dogs. One afternoon in 1903, they used a terrier mix—the pup that would serve as the muse for the statue. And so it was, that in an odd confluence of events, a statue created to symbolize everything wrong with science unintentionally memorialized a momentous scientific discovery. These two men fueled anti-animal experimentation activism, but they also helped to launch the inchoate field of endocrinology. The bronzed canine represented a real dog used in a classroom demonstration, one that was supposed to teach students about a new theory and a new scientific word: hormone.

Starling and Bayliss worked well together but were quite different. Starling had been raised in a working-class family, while Bayliss was rich. Starling had movie-star looks: thick blond hair, chiseled features, and piercing blue eyes. Bayliss looked like a vagabond, with bedraggled attire, a long narrow face, and a scruffy beard. (His son claimed he never shaved.) Starling was optimistic, outgoing, and impulsive. He thrived on results. Bayliss was cautious, introverted, and detail-oriented. He enjoyed the process. It's been said that Bayliss was so committed to his work that he initially declined an invitation to be knighted at Buckingham Palace because it conflicted with a physiology meeting. The two scientists were even related by marriage; Bayliss married Starling's sister, Gertrude, who was just as stunning as her brother. Starling married into money, wedding Florence Wooldridge, the wealthy widow of his former mentor, Leonard Wooldridge.

They were prominent physiologists long before their fateful hormone studies. They had investigated the heart, gathering evidence for what they later named Starling's Law, linking the organ's force of contraction to the force of expansion. They had explored how immune cells travel around the body. They had examined the wave-like propulsions that drive food through the intestines and dubbed it peristalsis—from *peri*, Greek for around, and *stalsis*, to squeeze.

Inspired by Ivan Pavlov, a colleague in Russia, the two physiologists switched from exploring the forces in the body to its secretions. That's what led to their endocrine work, the dog experiment and subsequent demonstration, and ultimately a court case. Starling and Bayliss wanted to test something Pavlov had recently theorized: that nerves chauffeured messages from the gut to the pancreas, triggering a release of chemicals.

On January 16, 1902, Starling and Bayliss conducted a terrifically simple experiment. After anesthetizing a dog, they picked away the nerves near the gut. Would the pancreas still release its digestive juice? If so, that meant the messages from the gut to the pancreas were ferried by something other than nerves. If the pancreas did not release its secretions, then Pavlov was right: messages traveled along nerves.

Bayliss and Starling fed the dog a lump of acidic mush to mimic digested food. Despite the lack of nerve connections, the pancreas churned out its juices. A mysterious chemical sent a signal to the pancreas, they concluded. Not a nerve.

Next, they cut out a piece of the dog's intestine and mixed it with acid. As before, that too simulated digested food. But this time, rather than keeping the intestine in its normal location, they injected the amalgam into a vein. The point was to put the mixture far from any nerves near the pancreas.

Victory! It worked just as they had hoped. They confirmed their initial experiment, and insisted that they, also, had isolated a specific pancreas-stimulating substance from the intestine. They declared that the process by which the pancreas released its juices had nothing to do with nerves but was rather a "chemical reflex." Starling named the intestinal secretion "secretin."

Secretin would one day be recognized as the first hormone to be isolated.

Pavlov then conducted an experiment similar to that of the British team. He, too, plucked away the nerves, aiming to confirm his initial speculation. But when the pancreas released its secretion, he concluded that he must have missed a few. The signals, he insisted, were traveling along lurking nerves, too small for him to see. Same study. Same results. Contrasting interpretations.

Pavlov, along with most researchers, could not surrender the long-held belief that signals within the body must travel along nerves, regardless of data showing otherwise. He had been right that the gut signaled the pancreas, wrong that it got from Point A to Point B solely via nerves, but nevertheless he was awarded the 1904 Nobel Prize in Physiology or Medicine for his research into digestion. He also made dogs salivate to the sound of a bell—the Pavlovian response, a discovery which earned him lasting name recognition though no prizes.

Starling and Bayliss announced their new ideas to colleagues at the Royal Society in 1902. They reported that they had "so far failed to obtain a secretory effect on the pancreas by stimulation of the vagi," the nerve that winds its way from throat to belly, adding that they were "therefore rather sceptical of the alleged presence in these nerves of secreto-motor fibres to the pancreas."

Skeptical of Pavlov? That was quite the rebuke to their respected Russian peer. Nerve transmission of chemical signals was an accepted theory. If not nerves, then what was it that transmitted messages? The other Society members couldn't comprehend that a mysterious chemical could deliver a message without running along the tracks of nerves. You might as well have told Paul Revere that one day he could alert the masses with email. The skeptics figured that there *must* be ever-so-tiny strands of nerves conveying messages, much the way factory workers pass widgets along an assembly line, hand to hand, in close connection. That kind of Industrial Revolution image was more in line with the way turn-of-the-twentieth-century scientists imagined things.

Pavlov was stunned that his notions had been discredited but accepted the British team's ideas graciously. "Of course, they are right," he reportedly said when he got wind of their assumption. "It is clear that we did not take out an exclusive patent for the discovery of truth." Still, he failed to mention in his Nobel speech that Starling and Bayliss had reshaped his theory.

As Bayliss clarified in an article in the *Lancet*, a medical journal, nerves didn't elicit the pancreatic secretion, nor did acid, as others previously suggested. "The secretion must therefore be due to some substance produced in the intestinal mucous membrane under the influence of the acid, and carried thence by the blood stream to the gland." In time, the debate would become moot, as scientists realized that it is not a question of nerves versus chemicals, but a complex push and pull among them that controls bodily responses. Even the salivary gland, known since the dog days of University College London to be triggered by nerves, has been shown recently to be swayed by hormones too. Some twenty-first-century studies, for instance, suggest that postmenopausal declines in estrogen and progesterone prompt dry mouth.

Bayliss and Starling presented their theory before the specialty called endocrinology existed. Their ideas were daring, bordering on foolhardy. They were dismantling dogma, debunking the nerve theory that had been around for decades. Looking back on their remarkable insights from a twenty-first-century perspective, Dr. Irvin Modlin, a Yale gastroenterologist, wrote that in one fell swoop, these two men created a discipline. What they described over a hundred years ago is still accepted today. Scientists know that secretin, a hormone, neutralizes acid pouring out of the stomach when food is digested; precisely, secretin staunches the stomach's release of gastric acid and promotes the pancreas's release of bicarbonate. In 2007, scientists discovered that secretin also regulates electrolytes passing in and out of the bloodstream. To put it simply, secretin is a hormone that aids digestion.

Starling and Bayliss knew that, despite the naysayers, they were onto a novel concept that could change the way scientists perceive the human body. For years, a small cadre of physicians had wondered about a chemical communication between distant body parts. For instance, doctors had noticed that when mothers begin nursing, the uterus contracts. The gut experiment provided some of the missing evidence. Or, as Bayliss told the Royal Society in 1902, "A chemical sympathy between different organs has often been assumed as e.g., between the uterus and the mammary glands, but we believe that this is the first case in which direct experimental proof has been afforded of such a relationship."

Their pivotal investigations were completed shortly before the

Royal Society announcement. But the demonstration that inspired the statue took place one year later, on February 2, 1903, when Bayliss used a dog to teach his theory to sixty University College London students.

Bayliss didn't know that two anti-animal experimentation activists had infiltrated his lecture. Lizzy Lind af Hageby and Leisa Katherina Schartau had moved to England from Sweden to enroll as part-time students in a nearby women's college. They wanted to learn a little physiology, but mostly they wanted to collect ammunition for the antivivisection movement. The women's college did not permit experiments on live animals; if a student wanted to witness one, she had to get permission from the professors at the men's college. As the women would later explain in court, they had enrolled as medical students in order to distinguish themselves from the vast majority of animal-rights activists, who protested from a platform of ignorance. They wanted to be able to speak the language of the scientists and use it against them.

The women were on the front lines of a movement that had been simmering since the middle of the nineteenth century, coincident with the rise of laboratory medicine. The more experiments that were conducted, the more scientists relied on dogs and cats. The more they relied on animals, the more they disturbed animal lovers. Thanks in part to the vociferous antivivisectionists, England became the first country to enact a law restricting animal experimentation. The 1876 Act to Amend the Law Relating to Cruelty to Animals (passed twenty-seven years before the professors' dog demonstration) stipulated three things: that only practitioners with special licenses could perform experiments on live animals, that an animal could only be used once, and that the

animal had to be given painkillers unless the drug would inter-
fere with the study. Antivivisectionists complained that the law
lacked bite.

The women came to University College London specifically to
cause a ruckus, but ended up being spectators at one of the most
pivotal endocrine demonstrations ever. To begin class that day,
Bayliss's assistant, Henry Dale, carried in the brown terrier mutt
and strapped him supine, legs splayed, on a black laboratory table
at the front of the lecture theater. They had chosen a dog that had
been used in a pancreas experiment a few months earlier—a choice
that would come back to haunt them in court.

Since the dog's pancreas had been mucked up, Bayliss focused
on the salivary gland. The point was the same: to demonstrate the
chemistry of the digestive tract. Bayliss leaned over the dog, slit its
throat, and peeled back the skin where the salivary gland wrapped
around the jawbone. He glided the knife down toward the dog's
Adam's apple. He snipped one of the threadlike lingual nerves connected to the salivary gland and fused a loose end to an electrode.
Zap. Buzz. Zap.

For nearly thirty minutes, the professor shocked the nerve. The
students peered closer. Nothing. Again. Zap. Buzz. Zap. Nothing.
As every experimenter knows, sometimes even the best-laid plans
go awry. The electrically stimulated nerve was supposed to jolt the
salivary gland to release its juices. These juices, or internal secretions, would activate the glands of digestion. The glands would
then do their job—stimulating digestion without traveling along
nerves. But nothing happened. Eventually, Bayliss nodded to Dale,
who carried the dog out of the classroom, removed the pancreas so
that it could be inspected by microscope to see if it had received
the chemical signals, and plunged a knife into the dog's heart to

end its misery. Later, Bayliss and Starling plumbed the pancreas for tiny nerves, hoping not to find any—which would support their chemical theory.

The class demonstration may have been a failure, because the salivary gland did not do what it was supposed to do, but it was just what Lind af Hageby and Schartau needed. They immediately began writing an antivivisection book, describing what they saw. They called it *Shambles of Science: Extracts from the Diary of Two Students of Physiology*. Giving a nod to the groundbreaking research conducted by Bayliss and Starling, they wrote that their intentions were "twofold, first to investigate the *modus operandi* of experiments on animals, and then to study deeply the principles and theories which underlie modern physiology." By "investigating the *modus operandi*," they meant gaining evidence that scientists were breaking the law on vivisection. They reported that they saw an open wound in the dog's belly, proof that it had been used in a prior experiment. Using the same animal a second time was unlawful.

Strike one against the vivisectors.

The women also saw the dog flinch, a sign that it was suffering pain. Laboratory animals must be given painkillers, according to the law.

Strike two.

They questioned where Bayliss and Starling had obtained the terrier in the first place. Rumor had it that scientists snatched dogs from their owners, that they scoured parks looking for runaway pets. "His master may have lost him early this morning," the women wrote, "but no advertisement and no rewards will bring this dog back." Such stories, whether fact or fabricated, added to the freakish aura of laboratory medicine.

The women also claimed that, during the lecture, Bayliss had reached into the pup, grabbed a piece of its intestine, and told the students that he had to be careful the entire thing didn't flop out. The male students, they claimed, chortled and clapped. They originally titled that chapter "Fun," but their publisher, an antivivisectionist crusader himself, demanded that they switch to a less sarcastic tone.

At the end of the semester, Lind af Hageby and Schartau handed their book, as well as their notes from every class, to Stephen Coleridge, a lawyer and president of the National Anti-Vivisection Society. That's when the dog statue racket began.

The women wanted Coleridge to sue the scientists, but he thought they'd have little chance in the courts. Judges tended to sympathize with the medical establishment. Also, animal abuse cases had to be filed within six months, and time was running out. Lastly, to file a suit, they would have to get the approval of a high-ranking legal administrator, who—like the judges—was known to side with scientists. In essence, Coleridge suggested they dodge the legal rigmarole. He had another idea: a demonstration.

Rather than work within the system, he suggested they appeal to the masses and get the public on their side. So, on May 1, 1903, Coleridge and his organization mobilized upward of 3,000 people to attend a speech at St. James's Church on Piccadilly, in the center of London. There, he waved the book *Shambles* and bellowed about animal abuses in science.

He called Bayliss and Starling's work a "cowardly, immoral and detestable" act. He read antivivisection testimonies from noted English writers, including Rudyard Kipling, Thomas Hardy, and Jerome K. Jerome. "If this is not torture, let Bayliss and his friends . . . tell us in heaven's name what torture is," he proclaimed.

The crowd hooted and hollered. A Battersea-based tabloid, the *Daily News*, reprinted Coleridge's speech word for word. The national press picked it up.

Bayliss, who shunned publicity, preferred to ignore the whole thing. But Starling ran on a shorter fuse and urged him to confront the mob that was making a mockery of serious science. Confident that the judiciary would be on their side, he convinced Bayliss to sue Coleridge for libel. Bayliss, hoping to avoid hullabaloo, first asked Coleridge to make a public apology. When Coleridge didn't respond, Bayliss went to the courts.

On November 11, 1903, students, vivisectionists, antivivisectionists, professors, scientists, and assorted activists lined up outside the Old Bailey courthouse. Some came to show support for the defendants, others for the scientists. The trial would not be about the morality or legality of animal experimentation; it was solely a case of libel. The plaintiff was the scientist. The defendant was the lawyer who had instigated the protest.

To Starling and Bayliss, it must have seemed that everything they had accomplished was being called into question. Colleagues doubted their theory of chemical secretions. The public challenged the way their experiments were conducted.

Starling, serving as a witness for Bayliss, admitted that the animal had been used twice but explained that since the dog was about to be killed, they had preferred to use it rather than experiment on another dog. Medical students, who came forth as witnesses, said the dog's twitching was a reflex, not an indication of insufficient painkillers. The trial lasted four days. On November 18, the jurors began their deliberation. It lasted for twenty-five minutes. They found Coleridge guilty of libel. The judge ordered him to pay £5,000, which amounts to more than half a million pounds sterling or three-quarters of a million U.S. dollars, in today's terms.

The medical students jumped out of their seats, yelling "Three cheers for Bayliss!" Bayliss donated the money to the physiology laboratory.

The *Daily News*, the working-class newspaper, called for a tightening of vivisection laws. "Here is an animal which worships and trusts mankind with an unreasoning fidelity," an editorial read. "Does not this overwhelming trust—this absolute confidence that glistens in the dog's eye—lay upon us some obligation?" *The Times*, which was known to side with the scientists, called the whole affair—which included women gaining entry to a medical lecture theater and Coleridge insulting eminent doctors—sneaky and reprehensible. The *Globe*, another British daily, lambasted Coleridge for "bringing vile charges against honourable men."

As for the students, the case inspired hooliganism. First they crashed suffragist meetings screaming, "Three cheers for Bayliss!" Chances were that the feminists were focusing on feminist causes, but the students conflated women's rights with animal rights. Any activism smacked of anti-Establishmentarianism, and that meant suffragists were probably antivivisectionists, too.

Two years after the trial, in 1905, Starling gave four weekly lectures at the Royal College of London. He presented his new theory, which emanated from his and Bayliss's experiments and also from studies done elsewhere in Europe and the United States. It was a theory of chemical control, rather than nervous control, over the body.

In his introductory speech on the evening of June 20, 1905, Starling summed up gland research, using the word "hormone" for the first time: "These chemical messengers, however, or 'hormones' (from ὁρμάω, I excite or arouse) as we may call them, have to be carried from the organs where they are produced to the organ which they affect . . ." It was said almost as an aside, yet the label stuck.

Starling explained what made these chemicals differ from other bodily secretions. These juices, he said, have to be "carried from the organ where they are produced to the organ which they affect, by means of the blood stream, and the continually recurring physiological needs of the organism must determine their repeated production and circulation through the body." This is a clear definition of hormones: they are substances secreted by a gland that target a distant site; they travel via the blood; they are crucial for the maintenance of the body; they are crucial to survival.

He proposed the same idea that had been demonstrated nearly half a century earlier by Arnold Berthold—the rooster-testicle-swapping doctor, who in pre-hormone days had recognized how the testicles worked but never publicized his findings as Starling had. Nor did Berthold realize he had hit upon something that related to every gland. Berthold suspected the idea of secretions hitting far-away targets, but was thinking only about testicles.

Starling went on to say that "internal secretion," the common parlance, didn't explain the phenomenon precisely. A secretion is just that: something that has leaked. He sought a term that described not just exudate but a chemical on a specific mission, a substance with a target and an ability to ignite a response. He turned to two friends, the Cambridge University classicists Sir William B. Hardy and William T. Vesey, who suggested something along the lines of *hormao*, Greek for "to arouse."

Other candidate names were suggested by Edward Schäfer, one of Starling's professors. He proposed "autocoid": "auto," from the Greek *auto*, "self," and *coid*, "cure"—our inner cures. That moniker, for whatever reason, didn't stick. A few years later, in 1913, Schäfer suggested using the word "hormone" only for internal chemicals that stimulate and the word "chalone" (from the Greek for "relax") for internal chemicals that inhibit. That didn't catch on either.*

And so, hormones became hormones.

In the first of his four talks, Starling said he suspected hormones were emitted by four glands: the pituitary, the adrenals, the pancreas, and the thymus. He avoided mentioning the testes and ovaries, because he didn't want his illustrious audience to think he was one of the quacks peddling testicle and ovary tonics to reverse aging. That was a money-making fad in the early twentieth century, with preparations made from various animal gonads touted to boost energy and libido, and reenergize just about anything that flagged with age.

In his second and third lectures, Starling asked his audience if the definition of a hormone should require criteria similar to those for a germ. When German researcher Robert Koch discovered germs twenty years earlier, he had insisted on a set of principles—or postulates, as he called them—which included the following: The proposed germ must be able to be isolated. It must trigger a specific disease when injected into a healthy organism (as, for instance, *mycobacterium tuberculosis* causes tuberculosis). It must always cause that specific disease, and nothing else; when retrieved from a

---

* Names were really important to Schäfer, who changed his surname to Sharpey-Schafer when he turned sixty-eight. He said it was to honor his professor, the famous William Sharpey, but others believed he wanted to sound more English and less German (he had grown up in England after his father, James William Henry Schäfer, moved the family from Germany). He also dropped the umlaut.

sick being and reinjected into another healthy being, it must trigger the disease.

Inspired by the germ pioneer, Starling suggested that a hormone is a hormone only if (a) removing a hormone-secreting gland results in sickness or death, and (b) implanting a healthy hormone-secreting gland yields relief. Sometimes a so-called hormone didn't fulfill the Starling criteria, but nonetheless it kept its status. Removing or damaging the pancreas, for instance, triggers diabetes—step one, accomplished—but unfortunately you can't cure a patient simply by implanting a new one. Step two, failed. And yet, the pancreas is still considered a hormone-secreting gland.

In his conclusion, Starling asserted that the more we learn about hormones, the more likely we will be to find cures for all sorts of maladies, from constipation to cancer. "An extended knowledge of the hormones and their modes of action," he said, "cannot fail to render important service in the attainment of that complete control of the bodily functions which is the goal of medical science." In a later talk, Starling would remark that his discovery "seems almost like a fairy tale." He predicted that scientists would one day unravel the chemical composition, synthesize hormones, and use them to gain control over our bodies.

Two years after the Royal Society lectures, on September 15, 1906, a typically rainy day, the dog statue was unveiled in Latchmere Gardens, a tiny patch of green in the middle of a housing development near Battersea Park. It was paid for by Louisa Woodward, a wealthy Londoner and antivivisection activist. A *New York Times* editorial called the inscription "outrageous" and a "mute testimony to antivivisection morals." It would remain intact for four years, despite the commotion and protests of 1907. In 1910, the mayor of the borough of Battersea asked Woodward to move *The*

The *Brown Dog* statue that stands today. *Courtesy of Jessica Baldwin.*

Brown Dog to her garden. She refused. On March 10 of that year, in the wee hours of the morning, a few policemen and four local workmen dragged the statue out of the garden and deposited it in a bike shed nearby. Then they smashed it to smithereens and melted it down. The *New York Times* predicted that the "statue or anything of its likeness will never be seen again."

The *New York Times* was wrong. In 1985, Geraldine James, an antivivisectionist and member of the still extant, though not thriving, Brown Dog Society, commissioned a second *Brown Dog* memorial. It sits today in a patch of roses in Battersea Park, well hidden from the crowds. If you want to see it, head to the north side of the park, past the joggers' path, past the fenced areas where dogs play. Surrounded on three sides by a low fence draped by thick leafy trees is a smaller version of the original. The new one has no water fountain, and the bronze pup no longer sits proud but rather cute, much to the chagrin of current animal rights crusaders.

Perhaps a few passersby will notice the inscription and will recall Starling and Bayliss not just for their animal experimentation but for their innovative ideas. These two men, an odd couple, were unifiers. Unintentionally, they united a disparate public that had become increasingly suspicious of scientific research into one angry group that pitted their ire at vivisection. They also united doctors from disparate fields, bringing together the adrenal doctors and the thyroid scientists and the pituitary researchers into one specialty that would be called endocrinology.

3.

Pickled Brains

TWO FLOORS BELOW the main reading area in Yale's medical school library is a room full of brains. Not the students: these brains are not housed in bodies. They're in jars. Some containers hold one brain, others a few slices. There are about five hundred, side by side, in glass cases that wrap the perimeter. In the center, there's another shelf of brains suspended from the ceiling over a long table surrounded by stools. You can study there if you're not distracted by the scenery.

The specimens were collected by Harvey Cushing, a pioneering neurosurgeon of the first decades of the twentieth century. When he operated on people with brain tumors, he'd save a piece of brain and tumor and stick it in a jar. Some had tiny tumors; some had massive ones. Often, he'd get the rest of the brain after the patient died. Cushing also asked fellow surgeons to donate to his heady corpus. Cushing's own brain is not part of the compilation; it was cremated with the rest of his body in 1939.

Cushing was a collector. He saved his meticulous medical records, which read more like mini-biographies, not like the tally of chemicals and blood tests in medical charts today. He saved drawings he made of his operations (he was an accomplished artist). And he saved before-and-after photographs of patients. Some postoperative photographs are now displayed next to the brain containers, showing patients with tumors bulging from their heads. When Cushing could not remove a tumor, he would remove a piece of skull so the lump could grow outward rather than compressing the brain. It was not a cure, but it relieved the patient of many of the agonizing symptoms.

Cushing also amassed boxes of correspondence from his illustrious colleagues—letters they wrote to him and also letters to one another. These missives offer a behind-the-scenes peek at medicine of the time—not just what doctors were doing to patients but also what they did to one another and what they gossiped about. They were best friends, at times, but also competitive. Some of the letters reveal dismay that administrators were attempting to transform their noble profession of healing into a profit-focused business. That was back in the first decades of the twentieth century. Cushing donated his valuable library of first-edition medical textbooks to Yale, too.

But the brains, marinating in formaldehyde and hidden for nearly half a century, are the most special and perhaps the most telling of Cushing's collections. They are souvenirs from Cushing's most audacious operations and scrupulous studies. For more than fifty years after his death, these specimens, along with their accompanying notes and photographs—a treasure trove of medical history—became a jumble of cracked jars, dusty records, and glass plates shoved into various crannies of the hospital and medical school. In the mid-1990s, they were discovered in the basement by a few drunken medical students. After a colossal effort of cleaning and organizing the material, the brains found their final resting place in the Cushing Center. The room, which holds about three-quarters of the collection, was specially designed for them and opened in June 2010. The non-restored jars (about 150, in various states of pickled disarray) remain in the basement of a student dormitory, waiting to be, if not quite revived . . . refreshed.

If you take it all in—the library brains, the basement ones, the notes, and the images—you step back in time to the very earliest days of brain hormone research, the days when Cushing proposed a theory that connected mind and body. Perhaps he was inspired by Starling's 1905 speech—the one in which he dubbed hormones "hormones"—to venture deeper. Before Cushing, the brand-new concept of a hormone provided a fresh way to explain the body. From Cushing's time onward, the field expanded to the brain.

Harvey Cushing was born on April 8, 1869, in Cleveland, Ohio, the youngest of ten children. His family was wealthy. His father, grandfather, and great-grandfather were physicians. The young Cushing was popular, smart, and athletic. He left the Midwest to attend Yale, then Harvard medical school, and after that Johns Hopkins to train in surgery. He married Kate Cromwell, a Cleveland girl who moved in the same country club circles. During his surgical training, he studied under William Halsted, who devised the radical mastectomy.

Cushing had to be the best at whatever he did. When he was a teenager, he was selected for a competitive Cleveland amateur baseball league and then played on the varsity team at Yale. His sketches of his surgical procedures were published in textbooks. He was also a skilled pianist. Years later, during a sabbatical from surgery and

Harvey Cushing's brain collection in the Cushing Brain Tumor Registry, housed in the Cushing Center, Cushing/Whitney Medical Library, Yale University. *Courtesy of Terry Dagradi, Yale University.*

research, he wrote a book about his mentor, Dr. William Osler, a founder of Johns Hopkins Hospital, which won the Pulitzer Prize in 1926. Amid all of this, Cushing suffered from bouts of depression.

Cushing devoted his life to his work, leaving little time for his five children. But he directed his wife as to how they should be raised. His sons were groomed to go to Yale. Both did, though neither graduated. After one son failed out, Cushing asked the dean of the medical school to accept the boy without a college degree, but the dean refused. The other son died in his junior year in a drunk-driving accident. Cushing's three daughters, known in the society pages as the Cushing Girls, were raised to marry well. They did—two times each. One married James Roosevelt, the son of President Franklin Delano Roosevelt, whom she divorced, and then married the millionaire and U.S. ambassador John Hay Whitney.

Another married William Vincent Astor, the heir to a $200 million fortune, whom she left for the painter James Whitney Foster. The youngest married Standard Oil heir Stanley Mortimer, Jr., whom she divorced to marry CBS founder William S. Paley.

Cushing was skilled, daring, and confident—three traits crucial to becoming a leading brain surgeon at that time, when most doctors didn't dare venture into the head. As his biographer Michael Bliss put it, "In the first decade of the twentieth century, Harvey Cushing became the father of effective neurosurgery. Ineffective neurosurgery had many fathers."

If you had a brain tumor, your best chance of surviving the operation was to have Cushing as your surgeon. By 1914, he boasted a mortality rate as low as 8 percent, compared with 38 percent among patients treated in Vienna and more than 50 percent in London. Mortality meant surviving the surgery, not the cancer, which usually killed the patient not long after.

Cushing's operating technique, like everything else he did, was meticulous, said Dennis Spencer, former chair of neurosurgery at Yale, who helped spearhead the Cushing brain restoration project. "Whatever approach he was going to use to get to a tumor, he had this incredibly good judgment in terms of where the tumor was, getting there without harming the brain, and then getting out." And he did it all without the benefits of modern accoutrements, such as ultrasounds and MRIs, that help doctors today locate tumors. Cushing also fine-tuned a way to relieve patients who suffered from trigeminal neuralgia, excruciating face pain triggered by damaged nerves, by prying apart the bundle of nerves that connects the face to the brain. (Nowadays, trigeminal neuralgia is treated with drugs, such as anticonvulsants, or radiation that numbs pain fibers.)

Beyond surgery, beyond writing, and beyond drawing (and beyond getting his daughters married to rich men), Cushing was fascinated by the burgeoning field of endocrinology and launched trailblazing hormone studies. Other surgeons may have read about the new hormone discoveries with amused curiosity, but Cushing found his own niche within this growing field. There was plenty of research into nearly all the hormone-secreting glands—the thyroid, ovaries, testes, parathyroid, and adrenals—but one gland remained a mystery: the pituitary. Cushing knew that it was neglected because no one could access it. No one, that is, except himself.

The pituitary dangles like an upside-down lollipop near the base of the brain. If you could poke your finger past the bridge of your nose all the way through your skull, you'd touch it. Along the way, you'd bang into the nerves behind your eyes, which explains why people with pituitary problems often suffer from vision problems: a pituitary growth can press against the eye nerves. The name "pituitary" comes from *pituata*, which means phlegm in Latin, because Galen, a third-century doctor, assumed that the gland's only job was to spit mucus. The pituitary, as even ancient doctors noted, was not just one sphere, but two adjacent lobes. The front one is called the anterior pituitary; the back one, the posterior pituitary.

In time, doctors would learn that these two lobes have different functions. Each secretes distinct hormones. They are like next-door neighbors who don't have much in common but close proximity. As a whole, though, the pituitary controls all the other glands in the body. For a while, it was known as the mother gland, until the 1930s, when scientists discovered that another organ in the brain, the hypothalamus, controls the pituitary. At that point, the hypothalamus got the mother-of-all-glands moniker.

When Cushing decided to explore this dangling pea-sized gland, very little was known about it. "The Chief's first and only true love," his secretary would call it. Within a few decades, he would be considered the leading expert on the pituitary.

Cushing explored the gland with the same audacity that he brought to brain surgery, doing things that others were too timid to attempt. When he suspected the pituitary released growth hormone—but before he had solid proof—he invited dwarfs to his clinic and fed them pituitaries extracted from cattle, to see whether the little people would grow. They didn't.

Cushing also attempted the world's first human-to-human pituitary transplant. In 1911, he extracted a pituitary from a baby shortly after it died and implanted it into a forty-eight-year-old man who had been diagnosed with a pituitary tumor. Newspapers heralded the experiment as a scientific breakthrough: "Broken Mind Cured," said the *Washington Post*. But the praise was premature. Six weeks after the operation, the recipient, William Bruckner, got sick again, with massive headaches and double vision. Cushing implanted another baby-brain gland. Bruckner died a month later.

Cushing would not admit that his transplant had failed. He claimed that the autopsy showed that Bruckner had died of pneumonia. He also blamed the obstetrician who had delayed two hours in delivering the baby's gland to the operating room.

Alongside his bold human experiments, Cushing performed animal studies. He began with the most fundamental question (can you live without a pituitary?) and culminated thirty years later with an exhaustive analysis of the cells that make up the gland. In the early days, he removed dogs' pituitaries, and also fed morsels of pituitary to other canines. He wanted to see what would happen if dogs had too little or too much pituitary gland, which meant too much or too few of the hormones it contained or controlled. His pituitary-free dogs died, so he concluded that one cannot live without this gland. (Doctors now know that dogs—and people—can survive without a pituitary, but they don't grow or mature; they are tired and have trouble burning calories. Nowadays, for those born with a malfunctioning pituitary, hormone treatment, to replace the missing hormones, is available.)

Unlike his predecessors, who fed chunks of whole pituitary to laboratory animals, Cushing tested the lobes separately. When he gave the dogs a piece of pituitary from the posterior lobe, blood pressure and urine flow increased, and the kidneys swelled. When he gave them a piece of pituitary from the anterior lobe, the dogs shrank to mere skeletons.

What was going on here? How did a little extra pituitary make such a huge difference? Did the pituitary control weight? Fluid regulation? Did the lobes communicate with each other, or were they separate entities hanging off the same stem?

Cushing was a keen observer. He noticed that his pituitary-free dogs didn't just fade away; they got sick in a peculiar way. Their bellies bulged. Their limbs atrophied. They got tired. Their ovaries or testes shrank. When he posed them on their hind legs, their physique was similar to that of many of his brain tumor patients, with skinny legs and swollen bellies. Could a pituitary malfunction explain all of this?

Cushing did what he did best: he collected. He asked colleagues to refer live patients to him. He also studied dead ones, trawling through morgues, cemeteries, and museums for the brains of people who were physically abnormal—too short, too tall, too fat. He measured the skull of a famous eighteenth-century giant displayed in the London Museum and found that the bones that cradle the pituitary were splayed, a clue that something was pushing on them—perhaps a pituitary tumor, which might have triggered his extraordinary size.* He sent one of his students to examine the pituitary of a recently deceased circus giant, and though the family refused to allow an autopsy the student slipped the undertaker fifty dollars to look the other way while he carved into the cadaver's skull. The student told Cushing that he saw splayed bones.

By 1912, Cushing had a compendium of notes on patients with suspected pituitary problems. He wrote about them and photographed them. He amassed case after case of men and women with the same fat bellies and skinny limbs as his pituitary-free dogs (doctors referred to it as a "lemon-on-toothpicks" look). These people had more than just a peculiar physique; they also sprouted hair in the wrong places, had slumped shoulders, and their skin was colored with bluish stripes. Their blood pressure was too high. The women had stopped menstruating. The men were impotent. They were exhausted, weak, and depressed, and suffered from pounding headaches. Nearly all were in their twenties. Many had worked as circus freaks before ending up in the hospital.

Cushing published his notes, and photographs of the patients naked, in *The Pituitary Body and Its Disorders* (1912). Cushing had detailed descriptions of his observations, but he couldn't prove that all of the subjects had tumors, so the book was a mix of evidence

---

* Cushing studied the skeleton of the Hunterian Museum giant, Charles Byrne. Byrne died after a career as a circus giant. He had begged that his body be dumped in the ocean so he would not have to play the freak for eternity. Nevertheless, his skeleton went to the Hunterian Museum where it has been for 250 years. Every now and then, activists and historians call for its removal, prompting a 2011 statement from the museum on the care of their human remains.

and conjecture. He claimed that some tumors or defects revved up the pituitary and others did the opposite, dampening the gland. He had three names for the ailments: hyperpituitarism, where the gland was on overdrive, as in the giants; hypopituitarism, where the patients were fat and tired; and dyspituitarism, which he posited was a combination of both. He concluded that some people had "polyglandular" syndrome: multiple glands gone awry. He envisioned a cascade of events, whereby a tiny brain tumor disgorged a substance that triggered the adrenals to pump out too many hormones, ultimately sending the whole body haywire. He identified the symptoms of this brain–body breakdown as weight gain, weakness, excess facial hair (particularly noted on women), and loss of libido.

In time, other scientists would name the adrenal hormone cortisol. It's a potent hormone that controls many bodily functions. Cortisol helps regulate blood pressure, metabolism, and the immune system. A spurt from the adrenal gland in the morning, doctors now know, maintains the body all day. Cortisol also helps promote labor, and coats fetal lungs so they can easily expand and deflate. But too much cortisol, as Cushing was beginning to elucidate, wreaks havoc on the body. In addition to the long list of ailments seen in his subjects, high levels of cortisol can trigger depression, psychosis, insomnia, heart palpitations, and brittle bones. Sustained high levels can kill.

Eventually, the ailment he described as polyglandular syndrome would be named for him: Cushing's syndrome and Cushing's disease. The difference between disease and syndrome depends upon where the problem is triggered: a pituitary tumor leads to Cushing's disease, while a problem arising in the adrenal glands leads to Cushing's syndrome. In both cases, the adrenals release too much cortisol, either because the pituitary emits a hormone that signals them to do so or because the adrenals themselves are faulty. The symptoms are the same: round, puffy face, fat belly with stretch marks, thin limbs, bone thinning, fatigue, and facial hair in women. These are the women that in the early days of the 1900s often ended up in the circus.

Years later, in the thick of a lecture tour about polyglandular syndrome, Cushing would write a scathing letter to the editors of *Time* magazine objecting to an article entitled "Uglies," about a Paris-based ugly contest. A woman didn't need to audition, and photos were submitted without the contestant's agreement. Entrants, according to *Time*, included a "fishmonger with warts," an "Italian Jew with erysipelas" (rash), a "pock-marked" taxi driver, and a Belgian nun. The point was to poke fun at beauty contests, or, as the reporters put it, "to offset the continental pestilence of beauty contests." But in Cushing's view, the effort to spare society from one superficiality only created another. These people needed doctors, not gawkers.

The article, which ran in May 1927, included a headshot of Mrs. Rosie Bevan (née Wilmot) positioned between a circus's Fat Lady and an Armless Wonder. The reporters found Bevan and splashed the image of a large-jawed woman with sagging eyes, a cropped helmet of hair, and sparse whiskers and beard. "This unfortunate woman has a story which is far from mirth-provoking," Cushing wrote, going on to say that Bevan probably suffered from acromegaly. "This cruel and deforming malady not only completely transforms the outward appearance of those whom it afflicts but is attended with great suffering and often with loss of vision," he wrote. He guessed that she had intolerable headaches and near blindness, and concluded, "Beauty is skin deep. Being a physician,

I do not like to feel that *Time* can be frivolous over the tragedies of disease."*

Cushing was a bold extrapolator. Based on his findings from severely sick patients, he advanced the idea that many people who were, if not extremely deformed, perhaps a little off—physically or emotionally—had one or two hormones out of sync. It was a whole new way of considering illness. It was, in fact, quite prescient.

Cushing also continued to fine-tune his theories of the pituitary. When he started out in 1901, he had a hazy picture, a speculation really, of how the pituitary controlled the body. He talked in terms of hyper (overactive) or hypo (underactive), but nothing specific. By the 1930s, as he was nearing retirement, he honed his ideas right down to the types of cells within this tiny gland. Speaking along the Eastern seaboard to audiences of leading experts at every major institution, he explained that the pituitary is not one homogenous organ. Within the anterior lobe, he said, there are three kinds of cells. An overgrowth of one type prompted enormous growth; too much of another type led to stunted sexual development.

Consider this: When Cushing was lecturing and writing his scientific articles, he was promoting a theory based on a yet undis-

---

* In 2006, Hallmark issued a satirical birthday card with poor Mrs. Bevan's photo again. It was sold in the U.K. with a joke about a British television show, *Cilla Black's Blind Date* (contestants choose a date unseen). The text on the card read, "When the screen went black, he was to always regret the words . . . I'll go for number three, Cilla." Like Cushing generations earlier, Dr. Wouter de Herder, a Dutch endocrinologist, saw the card while on holiday in Britain and complained to Hallmark, persuading the company to take the card off the market. On a pituitary tumor website, a blogger remarked that this story shows that while in the years since Cushing we've learned a lot about the diseases, "our attitudes to sufferers have hardly changed." For their part, Hallmark pulled the card and issued this statement, "Once we found that this lady was ill, rather than simply being ugly, then the card was . . . withdrawn immediately, as it would breach anything we would do in terms of taking the mick out of anyone who was poorly."

covered hormone and an entirely new concept about how the body worked, grounded on his notion that a tiny tumor had grown in a patient's brain. Sometimes doctors found a tumor on autopsy, but sometimes they didn't see one at all, despite probing about in the cadaver's head. Among the dozens of patients Cushing presented as proof, he had only found this particularly small tumor, which he called a basophil adenoma, in three of them.

In those days, if a doctor suspected a patient had a brain tumor, he X-rayed the head. The point wasn't to see the tumor (which wouldn't show up on X-ray) but to establish whether any bones were bulging, which would be circumstantial evidence of a mass. Cushing claimed that the basophil adenoma was so diminutive that bones didn't splay. In other words, his proof was lacking. Yet he believed that the tumor was there and that it pumped out a potent substance. He might as well have been trying to convince the audience that God didn't exist.

Now we know that he may have been right. Some tiny pituitary tumors are benign; they are small and slow-growing and don't spread to other parts of the body. With the advent of the sophisticated imaging tools that would come many years later, it might have been proven that some of Cushing's patients did indeed have tumors.

Cushing never doubted his assertions. Others did. A doctor at the Mayo Clinic in Rochester, Minnesota, dissected one thousand pituitaries from corpses and found basophil tumors in seventy-two that apparently showed no external symptoms. In other words, he claimed to have found growths in people without symptoms, debunking Cushing's theory. He called them not adenomas, as Cushing called them, but the snarky name "incidentalomas," suggesting that this was an incidental finding that had nothing to do with the symptoms Cushing had associated with them. Other doctors mocked Cushing by launching an Anti-Pituitary Tumor Club.

In a lecture at Johns Hopkins Hospital in 1932, Cushing said that endocrinology lent itself to the "temptation of impressionistic speculation." In other words, he didn't have as much evidence as he would have liked. "We are still groping blindly for an explanation," Cushing said, yet "out of this obscurity, those seriously interested in the subject have, step by step, been feeling their way in spite of pitfalls and stumbling blocks innumerable."

Today we know precisely what the pituitary does. The front lobe, which doctors call the anterior pituitary, discharges several hormones, including growth hormone and prolactin (best known for its role in milk production). It also fires so-called releasing hormones, which are hormones that prompt other glands to release hormones—messenger hormones of sorts. Gonadotropins, for instance, are hormones that direct the ovaries and testes to secrete estrogen and testosterone. The pituitary also releases thyroid-stimulating hormone, which signals the thyroid to release its hormone, and makes ACTH, adrenocorticotropic hormone, which prompts the adrenal gland to release stress hormone.

The lobe in the back, the posterior pituitary, makes vasopressin, which maintains fluid balance. It also spews oxytocin, which, among other things, makes the uterus contract during childbirth and the breast milk ducts squeeze afterward.

Cushing continued to operate, to conduct experiments, and to write more than 10,000 words a day until his chain-smoking got the better of him. By the time he was in his sixties, he could barely walk because of blood clots in his legs. He retired from Harvard in 1932, when he was sixty-three, and accepted a professorship at

Yale, bringing his assistant Louise Eisenhardt with him. She had been hired as his secretary in 1915, left four years later to get her medical degree at Tufts University (graduating top in her class), and returned to work for him as a neuropathologist. Cushing's debilitating moods and poor circulation stymied his plans to continue to operate; also, he was no longer dextrous. He spent most of his Yale days reading, teaching, and writing.

Cushing's vast brain collection was supposed to stay at Harvard, organized into a Cushing brain registry by Eisenhardt. But when Cushing felt that Harvard wasn't providing sufficient funding, he moved the entire stock to Yale. The jars arrived in New Haven in 1935. Cushing paid the equivalent of $100,000 by today's standards to have his patients' notes (all 50,000 pages of them) photographed and delivered to New Haven as well.

Eisenhardt remained Cushing's loyal partner as he continued to decline. He died of a heart attack on October 7, 1939, at the age of seventy.

That was the end of the Cushing era, but not of his brains.

Nearly thirty years after Cushing's death, Gil Solitaire, a neuropathologist, was hired by Yale. As he settled into his office, he opened a metal filing cabinet and found a jumble of jarred brains and empty whiskey bottles. Solitaire had a hunch that his office had once been the Cushing/Eisenhardt room, so he figured the brains and booze were Cushing's stash. Eisenhardt had been known to pound down a few at office parties.

Another Yale pathologist had been responsible for organizing the collection, but it never happened. The rest of the jars—the ones not in Solitaire's cabinets—were scattered around the pathology department. Eventually, they were moved to the basement of the medical students' dorm and forgotten about for decades.

In 1994, Chris Wahl, a first-year medical student, ventured to the dorm basement on a drunken dare and found the remarkable cache. "I think a few people in every class knew about them and I remember sitting at Mory's [a private eating club] with some upperclassmen and these guys saying you've gotta check out the brains," Wahl recalled. "We obviously weren't going to let that go unchecked so myself and about four or five other guys broke in down there. We ended up kicking in the bottom of the grated vent on the door and could reach in and unlock the door. And there was this room. I remember it vividly because we were a little afraid we were going to get in trouble and it was a creepy place and we were seeing brain specimens and there was a sign-up board near these creepy empty wine bottles of people who had gone down and seen the room."

A poster taped to the wall said "Brain Society" and was signed by students. If you could find the poster and sign your name as proof, you became a member. The society had an oath—"Leave your name, take only memories"—but no mission. Membership gave you bragging rights. For most students, going down there was just that, a been-there-done-that sort of thing. An initiation into a club few folks knew existed.

"I remember thinking this is creepy stuff, and then someone found the negatives, someone found them along the entire back wall, just shelf upon shelf upon shelf, floor to ceiling of these glass plate negatives in manila envelopes that were just brittle and you'd pick it up hold it up and they were incredibly chilling images of these people with brain tumors," Wahl recalled. "It was turn around and run scary."

The glass-plate negatives contained images of Cushing's patients before and after surgery. Some show huge bulges sticking out of their heads. Some are headshots. Some show the full body. Some patients are naked. Some clothed.

Tara Bruce, an obstetrician–gynecologist in Houston and former Yale medical student, remembered the brains. "It was a rite of passage," she said. She became a society member in 1994, which is to say that she signed her name on the poster in bold scrawl. "Everyone went to see the brains. It was surreal. I had just got to Yale and I remember thinking, 'I guess Yale has so much great stuff that they can shove a bunch of brains in the basement.'"

Wahl's claim to fame (before his current job as an orthopedic surgeon in Seattle and his previous one as head physician for the San Diego Chargers) was having been the one student among the cadre of drunken brain-chasers to do something with the stockpile. He had just finished a history of medicine class and had recently shadowed neurosurgeons, so it dawned on him that those jars might just be Cushing's collection. He went to see Dr. Dennis Spencer, the chairman of neurosurgery, and reported his hunch. In time, Wahl would write a thesis about the brains and—along with Spencer, a photographer, a medical technician, and an architect—spearhead the Cushing brain restoration project. That's how the brains transformed from medical detritus to medical museum.

Terry Dagradi, the Yale medical photographer and archivist of the collection, worked with a pathology technician to get the brains from the dorm basement to the morgue. It was a lot trickier than it had been in Cushing's day, when he was sending his brains from Harvard and collecting them from other doctors. In those days, they went by mail or were hand-carried on a train, just like any other package. But by the 1990s, when Yale launched the brain restoration project, the specimens were considered biohazards. Dagradi could not take the brains on public transportation without a special license. Just to take a brain across the street was enormously expensive. She and her colleagues devised a route that stayed on Yale property and avoided public roads, but that entailed loading the brains onto library carts and wending this way and that, up and down staircases, until they got from basement to morgue.

Nowadays, tours of the Cushing Center are available free to the general public. But if you are curious and can find a guide with a key, you may be able to see the brains that haven't been restored yet—the ones that are still in the basement. That's what I did with my fifteen students one spring afternoon in 2014. Accompanied by Dagradi, we traced Wahl's path into the basement, heading to the back of the massive building that houses the medical students' dorms, down a stoop and through a heavy metal door, stepping over some large pipes on the ground and ducking under some low-hanging ones, traipsing past large gated storage cages (one contained piles of sleeping bags, another held a mattress, another a bicycle, and another a headless plastic torso elucidating the belly organs). One cage contained a drum set and guitars; apparently some students held band practice down there. Eventually, we reached a thick green door, guarded, it seemed, by a large rubber garbage bin overflowing with sticky pads, the kind that catch rodents.

The vent that Wahl smashed has been replaced by a thick slab of wood nailed securely in place. The door is bolt-locked. A sign on it says, "Property of Neurosurgery."

Dagradi unlocked the door and a whoosh of formaldehyde washed over us. The room was dark and dank and dusty. Stalactites leached from the roof like white icicles.

Brains in old mason jars, hundreds of them, were stacked on old-fashioned metal library bookshelves that ran from floor to ceiling. Some specimens floated in formaldehyde. In others, the pre-

Unrestored portion of Cushing's brain collection in the basement of the dorms at Yale School of Medicine. *Courtesy of Terry Dagradi, Yale University.*

servative had evaporated through tiny cracks, so the bits and bobs of brain were puckered and withered. Some jars held only a few wisps of tissue. Others held a little chunk. A few held nearly half a brain. They were dated, most from the first few decades of the 1900s. Names were written on the jars. One jar contained an eyeball; another contained a fetus about an inch long. It was as if we had entered the lab of a mad scientist. Or a Disney movie about kids falling into a time warp and slipping into a spooky scientific experiment. Or worse, Hannibal Lecter's attic.

Drawers were filled with antiquated medical equipment, some used by Cushing to slice specimens. An old-fashioned metal gurney blocked one aisle. About eighty brains, Dagradi explained, were neither in the basement nor the library, but at the morgue where the cleanup took place. The jars en route to the morgue were in large white rubber vats on the floor, the kind that restaurants use for mayonnaise.

It felt as if the ghost of Cushing himself—a curmudgeonly, arrogant, tiny man with a beaky nose—was going to flit across the room and yell at us for trespassing. As we wandered up and down the aisle of brain bits, the silence of the place was broken by a loud swishing. Was it haunted?

"Someone must have flushed," said Dagradi—a reminder that we were below the student dorms. Or, from another perspective, that Yale medical students are spending their nights studying and sleeping on top of the foundations of modern endocrinology.

POSTSCRIPT

In the summer of 2017, doctors identified the genetic mutation that triggered a tumor in one of Cushing's patients who died more than a century ago. A thirty-four-year-old fisherman from Nova Scotia came to Cushing's Boston clinic in 1913 suffering from vomiting, irritability, profuse sweating, and numbness. "I grew all over," he told Cushing. The man had large hands and a jutting jaw. Cushing suspected a growth hormone–spewing pituitary and operated. When the patient died the following year, the autopsy revealed nodules in multiple glands. Fast-forward 104 years: Cynthia Tsay, a Yale medical student, working under the direction of Dr. Maya Lodish, a National Institute of Health endocrinologist, pried the brain from the glass case, removed a piece from the jar, and gathered the matching patient notes from Cushing's records. DNA analysis at the NIH revealed the precise genetic mutation and diagnosis: Carney complex, a syndrome named in 1985 that includes acromegaly along with multiple endocrine abnormalities. Lodish, intrigued by the brains ever since her days as a Yale medical student, has continued sleuthing other jars. Cushing, she added, was anti-Semitic and anti–women-in-medicine. "Here I'm this Jewish woman unleashing the brains from the glass and thinking he must be rolling over in his grave."

4.

Killer Hormones

O N MAY 21, 1924, two Chicago teenagers tried to get away with murder.

Nathan Leopold, or "Babe" as he was called, was nineteen. Richard "Dickie" Loeb was eighteen. They were both students at the University of Chicago, born and raised nearby in one of the most exclusive neighborhoods in town. That afternoon, they left their college campus, rented a car, and drove to the Harvard School, an elite private boys' school which they had both attended. Then they waited. The two had schemed for months and thought they had covered every angle to avoid suspicion.

They knew, for instance, not to drive Babe's red Willys-Knight car; that would be a definite giveaway. So they decided to rent a modest vehicle, choosing a blue one. They also lied to the Leopolds' chauffeur, telling him the brakes on the Willys-Knight needed fixing; that way, he wouldn't wonder about the rental. They rented the car under a false name, Morton D. Ballard. Their alibi—something to do with carousing all night with drunken girls—was rehearsed over and over to ensure they would have the same story, in case they were ever questioned. Babe and Dickie were brainy kids; they had each skipped grades and started college by the age of fifteen. But they were novice killers, so they weren't as thorough as they thought.

The boys had a shortlist of potential candidates, all sons of their parents' wealthy friends. They chose fourteen-year-old Bobby Franks because he was the last to leave school that day and was alone. They waited for him near the schoolyard and lured him into the car by offering him a ride home so he wouldn't have to walk. Then they drove a few blocks and bludgeoned him to death.

The corpse was found in the woods later that evening, with a pair of expensive horn-rimmed glasses lying nearby. The cops traced them to a high-end shop that had only sold three such pairs. One belonged to Babe Leopold.

Babe tried to explain it away as coincidence. He was an avid birdwatcher, he said, and had been in the same wooded area a few days before the body was dumped. The cops didn't buy it. Soon enough both boys confessed, each claiming the other was the ringleader.

The families hired famed defense attorney Clarence Darrow, the lawyer who would go on to defend John Scopes, the teacher sued by the state of Tennessee in 1925 for teaching evolution to public school students. For the Leopold–Loeb case, Darrow turned to science, too. His mission was not to prove the boys' innocence— they had pleaded guilty—but to get them life sentences instead of the death penalty.

The murder was quickly dubbed the "Crime of the Century." Newspapermen staked out the homes of the Leopolds and the

Loebs. They packed the courtroom. The case would, years later, inspire four films (one starring Orson Welles and another directed by Alfred Hitchcock), a few books (some fiction, some nonfiction), and one play. The driving question of the newspaper reports and the films and the novels, the question on everyone's mind, was: What had compelled these two boys who had it all—education, money, connections—to toss everything away for an afternoon of gruesome adventure? What was the motive?

The media stoked curiosity. Had the boys been emotionally neglected? It was reported that Babe's sickly mother had hired a flirtatious German governess to raise him. Dickie's mother was preoccupied with charity work, so he too had been fobbed off to a nanny, a painfully demanding one who punished him when his grades were less than perfect. During the trial, the public learned that the pair were occasional lovers and they both had a history of petty thievery. As a nine-year-old, Loeb pinched money from a lemonade stand he had with a friend. Leopold stole stamps from another boy's collection. Could these traits, the newspapers suggested, imply moral depravity?

None of these whys and wherefores—not the mothering or the sex or the pilfering—pulled all the puzzling pieces together. Yet, there was one theory that appealed to doctors, lawyers, and a lay public hungry for a scientific reason to explain deviant behavior, a newfangled notion gaining attention in medical journals and newspapers. The answer lay in the science of endocrinology.

Endocrinology exploded in the 1920s from an obscure science to one of the most popular specialties. Advice books touting endocrine cures abounded. Advertisements and feature stories in magazines added to the allure. With one discovery after another, hormones were deemed the cause of and cure for everything. The pituitary was shown to release hormones that stimulated the testes and ovaries. Estrogen was isolated and, soon after that, progesterone. Optimism skyrocketed when in 1922, at the University of Toronto, Dr. Frederick Banting, along with Charles Best, a medical student, saved the life of a fourteen-year-old diabetic with shots of insulin, ushering in a new generation of hormone therapy.

One year later, at a conference of the American Association for the Advancement of Science, Dr. Roy G. Hoskins summed up the enthusiasm for endocrinology this way: "When we see misshapen, stunted imbeciles transformed to normal, happy children, diabetics starving in the midst of plenty restored to health and strength, giants and dwarfs produced at will, sex manifestations engendered or reversed before our eyes by control of endocrine factors, who can regard endocrinology as other than a most significant phase of modern biology?" Hoskins was the president of the Association for the Study of Internal Secretions, which was founded in 1917 and would change its name in 1952 to the Endocrine Society, a leading professional organization.

If we could downgrade diabetes from a deadly disease to a chronic condition, thought the experts, just imagine all the other ailments we could fix! But murder? Was killing a disease? And if so, could lawbreakers be cured by hormone injections? Better yet, could hormone testing identify potential criminals even before they showed antisocial behavior? And, using the powers of hormone therapy, could we mold them into upstanding citizens?

It was a far-fetched idea in some regards, not so much in others. There was no proof whatsoever that a little extra of this hormone or too little of that one drove someone to kill. There wasn't even data confirming that a little too much of anything hormonal drove someone crazy, or drove them to anything at all. And yet, there was circumstantial evidence that hormones shaped behavior—a notion that had been around for centuries. The notion was based on trial and error— or nips and tucks—rather than on serious investigations. In the Ottoman Empire, for instance, men were castrated so that they would become sexless eunuchs to serve the royal court, a practice that tied the substances within the testes to personality traits. The scientific connection between internal secretions and temperament was first made in the early twentieth century, when in 1915 Walter Cannon, a Harvard professor, published *Bodily Changes in Pain, Hunger, Fear and Rage: An Account of Recent Researches into the Function of Emotional Excitement*. Cannon wrote that a sudden increase in the hormone adrenaline makes the heart pound and the breath come short and clipped. It resembled a panic attack, he said. His studies prompted other scientists to wonder whether other internal secretions affected emotions. "Here, then," wrote Cannon, is a "remarkable group of phenomena—a pair of glands stimulated in times of strong excitement and . . . a secretion given forth into the blood stream by these glands, which is capable of inducing by itself, or of augmenting, the nervous influences which induce the very changes in the viscera which accompany suffering and the major emotions."

The concept that hormones could infuse us with a killer instinct was a logical extension of Harvey Cushing's brain research. If out-of-whack secretions could make a woman grow a beard or a boy morph into a giant, as Cushing had demonstrated, then mightn't these internal substances turn a child prodigy into a violent criminal?

Cushing urged folks to sympathize with circus freaks because they were sick, not weird—but murderers press us beyond typical compassion. Bad guys may indeed have disturbed glands, but when there's a killer and a corpse, are we to believe they are both the victims? Or, as a *New York Times* reporter put it, referring to the biblical story, "It is possible that Cain's endocrine organs were functioning improperly and that he was as much a victim as his brother." This was the troubling part of the hormone-crime theory. It may have had scientific merit, but what were moral people to do with the information once the crime was committed? Should there be leniency because a killer's hormones went haywire?

For doctors, the insights offered a new way to think about the human condition. People were no longer just a jumble of nerve connections. In the 1920s, people became their hormones. Hormones were *us*.

The hormone-crime theory wasn't a switch in thinking but a unifying concept. Hormones affected the nerves in our brain, which in turn swayed our subconscious desires. "Accumulating information during the past fifty years has pointed to an importance of the endocrine glands for the problems of the science of psychology," explained Dr. Louis Berman in the austere journal *Science*. "I will propose the word 'psycho-endocrinology' as the name for that branch of science which deals with the relations of the endocrine glands to mental activities, as well as to behavior, including the individual characteristics in health and disease, summarized in the term personality."

Louis Berman had medical credentials and marketing savvy. If he had been around in the twenty-first century instead of the twentieth, he'd have his own TV show. He was an associate professor at Columbia University, authored some forty scientific articles, and was a member of several elite medical organizations including the New York Endocrinological Society, the American Medical Association, the American Association for the Advancement of Science, and the American Therapeutic Society. He was also a director of the National Crime Prevention Institute. A respected researcher, he isolated a hormone from the parathyroids, four tiny glands in the neck, and investigated its relation to calcium balance. He called the hormone parathyrin. Today it's called parathyroid hormone, or PTH, and is known to control calcium levels in the body.

Berman had a thriving Park Avenue practice, where he hobnobbed with the literati. Ezra Pound and James Joyce were endocrine patients as well as friends. "My dear Rabbi Ben Ezra," Berman wrote to Pound, using a nickname given to the poet by fellow poet Robert Browning, after a poem of the same name. They corresponded about their travels and gossiped about Joyce, the Irish novelist. Berman wanted to treat Joyce's daughter, Lucia, with hormones for her depression. "I don't know whether you've heard of the new insulin treatment for dementia praecox which I hear has also been used successfully," he wrote, adding, "another great triumph for endocrinology." (Dementia praecox was medical jargon for madness.) Berman tailored diets to balance his patients' hormones.

Berman was also an audacious extrapolator, sprinkling facts and then some into the books on health he wrote for the general public. He said that some people had too much output from their adrenal glands, which made them excitable and manly. Others had too little, with the opposite result. The "adrenal centered," he wrote in *The Glands Regulating Personality*, will have high blood pressure and masculinoid traits. The "adrenal inferior" will have low blood pressure and suffer from a constant weakness and fragility. Women who do not menstruate regularly do not have the right balance of female hormones, he asserted, and "will also be aggressive, dominating, even enterprising and pioneering, in short masculinized ovaries."

Yet Berman's books sold well and appealed to a broad public.

Berman simplified what other doctors couldn't and gave readers optimism with straightforward (though unproven) hormone remedies. He proclaimed that hormones would cure crime, insanity, constipation, and obesity. They would foster a better society, he forecasted. It would no longer be *survival* of the fittest; endocrinology would change us all into the fittest. Indeed, he predicted a planet of superhumans: the "ideal normal," he dubbed them. "We will be able to govern man's capacities in every detail so that we can create the ideal human being," he said. "The problem will be the selection of the 'ideal type'"—as he saw it, a sixteen-foot genius who didn't need to sleep.

Berman's ideas struck a chord in the 1920s, partly because the country wanted to figure out a way to deal with a nationwide surge in crime. Despite the flappers and the speakeasies and the Great Gatsbyish parties, there was a feeling that vandalism and murder were surging. The Ku Klux Klan peaked in popularity. Gangsters flourished. The exploits of Al Capone, the gang lord of Chicago, Bonnie and Clyde, the husband-and-wife bank robbers, and John Dillinger, another bank-robbing gangster, were splashed across newspapers in banner headlines, along with the story of Nathan Leopold and Richard Loeb.

By assessing a person's hormones, Berman claimed, it was possible to identify those likely to commit violence. Berman maintained that he could assess one's hormone "type," or rather predominance, by studying the face. Is this an ovary type of person? An adrenal type? A pituitary type? In effect, he was suggesting that personality is shaped by one minuscule gland. And, Berman said, he could use his evaluation to predict a person's future. Was he or she destined for leadership? For popularity? Berman's books listed the putative hormonal makeup of various luminaries. He worked backward, as their pathways to success or failure were already clear, but Berman claimed that their lives had been predetermined by their hormonal makeup. Napoleon and Abraham Lincoln were pituitary types. Oscar Wilde, thymus-centered. Florence Nightingale, a thyroid-pituitary mix.

Berman wasn't the only one promoting gland medicine as cure-alls. So-called organotherapy—remedies made from grinding up organs—was big business during the 1920s. Thyroid was used for myxedema (the medical term for underactive thyroid), pancreas for diabetes, kidneys for urinary tract ailments. In 1924, G. W. Carnrick Company, the makers of a slew of endocrine products, published a pamphlet that included remedies for 116 supposedly hormone-based ailments. They claimed that their brand of epinephrine suppositories (also called adrenaline) worked for hemorrhoids, vomiting, and seasickness; whole pituitaries eased headaches and constipation; and testes cured sexual neuroses. Testicle extracts were sold to cure epilepsy, weakness, cholera, tuberculosis, and asthma. "We are the creatures of these glands," said another endocrine doctor. Glands "are not only the arbiters of reactions and emotions, but," he added, "they actually control character and temperaments, whether for good for ill."

Criminal impulses could be explained by another set of hormones gone awry, said Berman. "Thyroxin, parathyroid, adrenalin, cortin, the thymus hormones, the gonadal or sex hormones, the pituitary hormones, the pineal hormones, all have their fundamental effect upon the statics and dynamics of the personality through their effect upon the nervous system," he wrote in an article for the *American Journal of Psychiatry*. In other words, hormones could drive a man to kill.

Berman's unscientific stretching of the truth rankled his colleagues, who questioned his professional integrity. In the *International Journal of Ethics*, a reviewer wrote that Berman's book "should be taken with a considerable dose of skepticism." In the *American Sociological Review*, another critic called it a "mixture of fact, half-truth, surmise, speculation, and hope—which is neither good science, good art, nor even good entertainment." Yet he maintained a wide net of advocates. Margaret Sanger, the birth control advocate, was among his fans. "For a clear and illuminating account of the creative and dynamic power of the endocrine glands, the layman is referred to a recently published book by Dr. Louis Berman," she wrote.

He did not convince H. L. Mencken, the editor of the *American Mercury*. "Every truth needs men who patiently plod, year after year, to formulate their hypotheses," wrote Mencken. "Berman is not one of these. But the new truth also needs trumpeters. One must forgive his inclinations to make too loud a noise and to introduce variations uncalled for by theme, to reveal his own versatility."

Serious academics cringed at doctors who preened for publicity. Or perhaps they were irked that one of their own was doing it better than they could themselves, and wished they were more like him. Dr. Benjamin Harrow, in his popular 1922 book *Glands in Health and Disease*, credited several Columbia University colleagues but failed to mention Berman. He alluded to Berman's work as "fact mixed with fancy," adding that "imagination, not sufficiently tempered by self-criticism, is apt to enlarge a molehill into a mountain."

There were also believers among the medical establishment. In 1921, at the Second International Congress on Eugenics, held at New York City's American Museum of Natural History, Dr. Charles Davenport, a eugenics crusader, kicked off the conference with a lecture that was in part about hormones and their impact on deviant behavior. The following day, in a session dedicated to gland research, Dr. William Sadler told colleagues that "severe disturbance in the endocrine system unfailingly leads to more or less marked criminalistic, immoral, and antisocial behavior."

Eugenics, which was popular among policy-makers, promoted the breeding of so-called good people with one another, as one would breed champion dogs. Eugenicists also called for the sterilization of those considered too stupid, deformed, or otherwise unfit to procreate. The Supreme Court sympathized with the cause. In the 1927 ruling *Buck v. Bell*, Justice Oliver Wendell Holmes, Jr., wrote that permitting compulsory sterilization of the unfit and the "intellectually disabled" was necessary "for the protection of the health of the state."

Berman pointed out that eugenics was a messy science, because smart, fit parents are not guaranteed to produce smart, fit children. He argued that working with internal secretions offered a surefire way to promote a healthy society. "We may now look forward to a real future for mankind because we have now before us the beginnings of chemistry of human nature," he wrote in his blockbuster book *The Glands Regulating Personality*. He called for a nationwide program to assess schoolchildren for their endocrine status. They could then be treated with some hormones to boost good qualities and other hormones to quash bad ones. For Berman, endocrinology was religion. As he put it in his 1927 book *The Religion Called Behaviorism*, "Christianity is dead. Judaism is dead, Mohammedanism is dead, Buddhism is dead, for all spiritual purposes. Slowly but steadily a new, powerful religion is growing into maturity in the United States as a result of a new psychological movement. It calls itself behavioralism. The body, the soul, human nature operates by way of chemicals, internal secretions or glandular effects." In hindsight, it's easy to see where Berman stretched the truth. Nearly a century later, it's hard, if not impossible, to know if he was truly a believer or a charlatan, or if his readers really bought into Berman's theories.

In 1928, Berman began a three-year investigation of 250 juvenile delinquents and criminals in Sing-Sing prison in Ossining, New York. He drew blood, measured metabolic rate, and X-rayed various parts of their bodies. Comparing his results to a group of healthy controls taken from among the ordinary population, Berman concluded that criminals have more than three times as many endocrine disturbances as law-abiding citizens. Murderers, he said, had too much thymus and adrenal hormones and not enough parathyroid hormones. Rapists: too much thyroid and sex hormones, not enough pituitary. Muggers and beaters: low on gonads (ovaries or testicles), high in adrenal. He did the same for those convicted of fraud and arson, putting each group of criminals into a tidy box with their own unique chemically disturbed makeup. He presented his findings in a talk at the New York Academy of Medicine in 1931 and published the results in the *American Journal of Psychiatry* the following year. The lengthy article was filled with data and charts showing the national increase in crime and the cost of crime, but it was thin on methodology. Nevertheless, Berman concluded that his work should be the basis for preventive medicine. "Every criminal should be examined for the presence of signs of endocrine deficiency and imbalance," Berman wrote, "including the pituitary, the thyroid, the parathyroids, the thymus, the adrenals, and the gonads, as part of the general examination to be added to psychiatric and social data."

For nearly half a century, defense attorneys had invited psychiatrists into the courtroom, hoping for scientific evidence that would get their clients off the hook. Even though Berman's bestsellers weren't published until after the Leopold–Loeb trial, his ideas were being widely circulated around that time and were debated among doctors. For the lawyers in the Leopold–Loeb case, Berman's psycho-endocrinology offered a new strategy. Clarence Darrow hired two experts in endocrinology: Dr. Karl Bowman, chief medical director of the Boston Psychopathic Hospital, and Dr. Harold Hulbert, a neurologist at the University of Illinois. Both were interested in the impact of hormones on the brain.

On June 13, 1924, the doctors met the two killers in a stark room inside the prison and began their examinations. A gaggle of reporters with binoculars clustered behind bushes on the far side of the courtyard, hungry for a visual detail to spice up their daily Leopold–Loeb updates. The doctors brought medical paraphernalia—an X-ray machine, a sphygmomanometer (to measure blood pressure), and a metabolimeter (to measure metabolism). The metabolimeter was an early-twentieth-century device that had a metal canister on a knee-high pole and tubes dangling off it. An oxygen tank pumped air into one of the tubes and a patient sucked air from the other one. Using a formula that included the patient's weight, height, and the time it took him to inhale, doctors got a number that purportedly told them the rate at which calories were being burned—in other words, the metabolic rate. This, they claimed, was a measure of hormone health. (Today we know that metabolic rate is not a way to assess overall hormone health, though it can provide a clue to the functioning of thyroid hormone, which is connected to metabolism.)

The two doctors used images from the X-ray machine to glean clues about hormones, too. Never mind that the images showed bones, not glands: if a gland was too big, the thinking went, it would

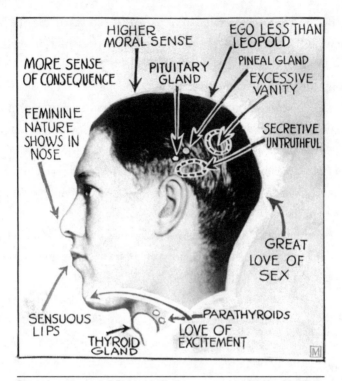

HIGHER
MORAL SENSE

EGO LESS THAN
LEOPOLD

MORE SENSE
OF CONSEQUENCE

PITUITARY
GLAND

PINEAL GLAND

EXCESSIVE
VANITY

FEMININE
NATURE
SHOWS IN
NOSE

SECRETIVE
UNTRUTHFUL

GREAT
LOVE OF
SEX

SENSUOUS
LIPS

THYROID
GLAND

PARATHYROIDS

LOVE OF
EXCITEMENT

Phrenology diagram of Richard Loeb that appeared in the *New York Daily News*. *New York Daily News Archive/New York Daily News/Getty Images.*

push against a bone. So if you saw a bone shoved aside, you could be pretty sure a gland was to blame. That's how Cushing, a few years earlier, had studied the pituitary, by seeing if it splayed bones in the brain. "These are methods that are now more or less standardized and in use in endocrine clinics and researches throughout the world," Berman wrote.

The examinations of Leopold and Loeb, which included the physical assessments and extensive psychiatric interviews, lasted nineteen hours and stretched over eight days. The result was a 300-page, 80,000-word report.

Before the psychiatrists were called to the stand to talk hormones, other doctors—Freudian analysts—presented their testimony for the defense. One described Babe Leopold as a short, skinny kid with a sickly complexion who had a stellar academic record. He studied Nietzsche, birds, and pornography. He spoke, it was said, eleven languages. He didn't have a lot of friends, but he adored Dickie Loeb. They had sex on occasion. Dickie, in turn, was described as a blond, blue-eyed, fetching fraternity boy. Unlike Babe, he swarmed with friends of both genders. Women continued to woo him long after he was charged with murder. The psychiatrists categorized him as having a normal intellect; he was not a brainiac like his friend. Yet he had "infantile emotional characteristics."

On August 8, Harold Hulbert, the endocrinology expert for the defense, walked to the witness box burdened with stacks of papers and loose-leaf binders. He seemed nervous and young compared to the gray-haired, self-confident physicians who had preceded him. Hulbert kept looking at the notes stacked on his lap and did not make eye contact with the prosecuting attorney, despite all the coaching he had received from Darrow's team. The prosecutor attacked the testimony of the Freudian analysts, claiming it was based on hearsay from criminals who were likely to lie. In rebuttal, Hulbert attempted to show that his hormone analysis offered hard, indisputable evidence.

There was one problem: while the data may have been indisputable, the interpretation was up for grabs. That's often the case. The way scientists generate theories based on evidence isn't always straightforward. They are influenced by their own preconceived notions of health and illness, by what seems to make sense at the time. That's how they make strides in knowledge—but it's also how they can go astray. Sometimes other researchers distinguish fact from fancy years later. Sometimes they never know.

Cushing took a leap when he shaped his theory of ever-so-tiny tumors in the brain wreaking havoc on the body. It turns out that he was right. But some of his data wasn't interpreted properly; years later, some experts would say that many of his patients didn't have brain tumors after all. As is always the case, only hindsight will tell whether a scientist was a trailblazer who carved a new path or a well-meaning investigator who hung a left at the wrong intersection.

The Bowman–Hulbert report, among other things, concluded that Dickie Loeb, the popular ringleader, suffered from multi-gland syndrome. His metabolic rate was minus 17 percent, a sign, they claimed, of glandular dysfunction. Babe Leopold's metabolic rate was minus 5, abnormal but not terribly so—but X-rays revealed severe brain damage. His sella turcica, the part of the skull that holds the pituitary, was slammed shut. Worse, his pineal gland was calcified.

"Nathan Leopold," said Hulbert, "seems to be a case of definite endocrine disorder involving particularly the pineal and the pituitary glands and the autonomic segment of the vegetative nervous system, associated with a cardio-vasculo-renal inferiority."

The pineal is a pea-sized gland shaped like a pinecone that sits deep within the brain. It calcifies with age. Leopold's had hardened way too early, according to the doctors. Descartes called the pineal the seat of the soul. Madame Helena Blavatsky, a founder of new age philosophy in the early 1900s, considered the pineal gland the "third eye," a notion that lingers among some yoga enthusiasts. Today, we know the pineal gland emits bursts of melatonin that control our circadian rhythm, our internal clock. In Leopold and Loeb's time, it was connected ever so tentatively to sex and intellect. The doctor explained that because of Babe's hard pineal, he had too much libido, even for a nineteen-year-old boy.

Hulbert gave Darrow precisely what he wanted: expert testimony asserting that the boys were behaving under the influence—under the pall, that is—of their deeply disturbed glands. As Hulbert added, these glandular defects "remove the ordinary restraint which individuals impose on themselves." After several days of questioning, Hulbert reiterated to the prosecutor that the "summary of the psychiatric findings in Richard Loeb based on my study is that he has endocrine disease which keeps him adolescent . . . and these delinquencies, including the Franks case, are the ultimate product of all the previously mentioned things."

It would be up to the liberal Judge John R. Caverly, not the psychiatrists or the endocrinologists or the lawyers, to decide whether these scientific claims held up in court of a law. Because the boys had pleaded guilty, this was not a trial by jury but a presentation of arguments to the judge who would decide their fate. As Hal Higdon, author of *Leopold and Loeb: The Crime of the Century*, wrote, "the so-called trial of the century therefore would not be a trial."

On September 10, 1924, at 9:30 a.m., some two hundred people mobbed the courtroom—the boys' families, the lawyers, and reporters from across the country. Everyone else in Chicago stopped whatever they were doing to congregate around radios tuned to WGN, which was broadcasting the decision live. Judge Caverly recognized that the doctors' careful analysis contributed to the field of criminology and that the value of their reports lay "in their applicability to crime and criminals in general." However, he said, "the court is satisfied that his judgment of the present case is not affected thereby." Put more simply, he said that even if the endocrine connection to crime was crystal-clear, even if the boys' hormones had dictated their behavior, that didn't mean they could get away with murder.

For the crime of murder, he sentenced both boys to Joliet Prison in Illinois for their natural life. Because of their young age, the judge was persuaded not to impose the death penalty. For the charge of kidnapping, he also sentenced them both to ninety-nine years.[*]

Nine years later, on January 28, 1936, Dickie Loeb was slashed to death with a razor by a fellow prisoner, James Day. Day claimed he was defending himself against Loeb's sexual advances. Babe Leopold was released on parole after serving thirty-four years, during which he was deemed a model inmate. On February 5, 1958, he moved to Puerto Rico, where he became a medical technician and married Trudi Feldman, a doctor's widow. He died of a heart attack on August 29, 1971, at the age of sixty-six. His body was donated to the University of Puerto Rico for no particular purpose. Perhaps it was peeled apart by medical students in a first-year anatomy class. There is no mention of anyone studying his glands.

---

[*] Joliet prison, opened in 1858 and shut in 2002, was used in the opening scene of the 1980 *Blues Brothers* movie and was also the location for Fox Network's television show *Prison Break*, which ran from 2005 to 2009, and the 2006 comedy film *Let's Go to Prison*.

5.

The Virile Vasectomy

Louis Berman, the psycho-endocrinology doctor, had big ideas: he wanted to use hormones to make the world a better place. A nation of chemically balanced bodies equaled a well-adjusted society, one free of crime, obesity, stupidity, and all the other traits that Berman linked to defective hormones. Utopia, brought to you by hormone specialists.

Eugen Steinach, a physiologist from Vienna, had big ideas too, but of a different sort. Whereas Berman considered the big picture, Steinach focused on the small one—one man at a time. Beginning in the 1920s, and for nearly twenty years, Steinach pioneered one of the most popular and controversial rejuvenation treatments. He claimed that vasectomies boosted sex drive, intellect, energy, and just about anything else that withered with age. Steinach believed that blocking the exit of manly juices (which is what a vasectomy does) prompted a congestion of them, much the way a traffic jam causes a pile-up of cars.

If you rate success by the quantity and quality of scientific evidence, vasectomies for rejuvenation don't rank high. If, on the other hand, you rate success by testimonials plus the number of paying customers, the practice was a global sensation. It was so popular, in fact, that Steinach's name became a verb: to Steinach meant to do a rejuvenating vasectomy. Sigmund Freud was Steinached. William Butler Yeats, the poet, was Steinached.

Steinach wasn't Steinached. Maybe that's why he didn't look rejuvenated. By the time he was promoting his treatment, he looked like an old man, with a long gray beard and a handlebar mustache. He wore stern dark suits more suitable for a mortician.

And Steinach never Steinached anyone. That's because even though he was a doctor, he didn't have patients. He preferred to do research on laboratory rats, and then instruct his surgeon friends how to cut people in the same way he clipped rodents. Steinach said that results were only guaranteed when he supervised, and he must have observed hundreds of operations. At least thousands of additional men were Steinached without Steinach in the operating room.

The 1920s was, to be sure, an exciting time for endocrinology, but it was also a messy one. Discoveries flourished in the same profusion as quack remedies. Hucksters and serious scientists were all mucking about in the same pool of theories and often coming up with nearly the same results. On either side of the divide, there were nostrums and diets and questionable procedures alleged to fix this ailment or that one. For the consumer, it was often tricky to distinguish swindlers from specialists. You could say the righteous practitioners were believers; if things didn't go as planned, it was a well-intentioned mistake. They were doctors and members of the established elite. The outright con men, on the other hand, were the guys who were in it just for the money, pushing cures they knew didn't work. But there was a vast gray area in between. And really, who's to know a man's intentions? It's not always easy to know whose motives were unsavory and who was swept away by the enthusiasm of the time.

There was the respectable Serge Voronoff, a doctor in Paris, who transplanted ape testicles into men to boost their virility. The medical establishment considered him a well-meaning but mistaken surgeon. Then there was the fabulously shameful John Brinkley, known as the Goat Gland Doctor because he peddled goat testicles to enhance human sex drive. He made a fortune. Clients shopped for their favorite pair on his farm and underwent surgery in his kitchen, with Brinkley's wife assisting. Brinkley wasn't a real doctor; he had bought a medical degree in Italy.

The medical elite worried about the effect of all of this quackery on the status of medicine. "It is pathetic if not disgusting to witness this endocrine orgy now rampant in our profession," Hans Lisser, a San Francisco–based endocrinologist, wrote to Harvey Cushing in 1921. "Much of it the result of abysmal chaotic nonsensical ignorance, much of it alas the result of commercial greed. Endocrinology is fast becoming a mockery and a disreputable business, and it is high time that some honest fearless words were uttered."

Like the Parisian Voronoff, Steinach was considered a proper scientist, not a quack. He was nominated eleven times for the Nobel Prize (not because of his vasectomy cure but for legitimate research into sex hormones), directed one of the foremost laboratories in Europe (the physiological department of the Biological Institute of the Academy of Sciences in Vienna), and published some fifty scientific articles. Among his many advances to the field, he discovered that cells lining the sperm duct (called Leydig or interstitial cells) produce testosterone.

Steinach's libido-boosting vasectomy was based on theories that had been bandied about for centuries. Since ancient times, healers have mashed animal testicles and ovaries into potions, desiccated them into powders, and dissolved them in medicinal cocktails or blended them with food. In 1889, seventy-two-year-old Charles Édouard Brown-Séquard, a Paris-based neurologist, injected himself with a slurry of testicle secretions from guinea pigs and dogs and claimed that it boosted his libido, increased his strength, quadrupled the length of his urine stream, and regulated his bowels. He felt thirty years younger, too. Steinach felt his approach was more scientific than that of Brown-Séquard. Brown-Séquard announced his discovery on June 1 of that year, a date he considered the birth of the science of endocrinology. Not everyone agreed. Many colleagues wondered how someone who had made great strides in a serious medical specialty could have gone so far off course. The press mocked him. A German medical journal said Brown-Séquard's "fantastic experiments with testicular extracts must be regarded almost as senile aberrations." Another scientist wrote that his lecture "must be regarded as further proof for the necessity of retiring professors who have attained threescore years and ten."

Nevertheless, Brown-Séquard's testicle-juice shots were *the* treatment du jour, sought out by men who wanted to brush away a few years of aging. At least, they were the rage for about five years, until Brown-Séquard died of a stroke at the age of seventy-six, a death that was fitting for a man his age then but was considered pretty young for someone supposedly rejuvenated. Not surprisingly, his demise was a crushing blow to his cure.

Steinach contended that his vasectomy reactivation technique trumped prior cures because it was risk-free and all-natural. He said the twenty-minute operation (a cut and suture of the sperm duct) was completely safe. Tweaking innate hormones, he added, was preferable to implanting foreign ones.

Men lined up for vasectomies, believing that they would become stronger, wiser and sexier. Yeats said it "revived my creative power, it revived my sexual desire, and in all likelihood will last me until I die." A sixty-one-year-old man (one of many testimonials in Steinach's memoir), who had been tired and sad and had lost all interest in sex claimed that after the operation "my memory is better, I understand things quicker. I live now like a man about 40 or 50 years old, and am in such good spirits that I find myself going about singing."

The rise and fall of the revitalizing vasectomy is a demonstration of the powers of placebo and publicity—how, even in medicine, being in the right place at the right time can make all the difference between flop and phenomenon. Steinach tapped into a society that was willing to try new hormone therapies, eager for self-improvement techniques, and able to pay for them.

In the United States and Europe, the years between the world wars were a time to turn inward, away from global affairs. Self-help books sold wildly and self-proclaimed healing gurus flourished. If you could afford it, you were lying on a couch being analyzed by a Freudian-trained psychologist. Women bought diet books to starve themselves into straight-hipped flapper dresses. Men read muscle magazines to learn tips from the brawny exercise mavens, such as Bernarr Macfadden, a Charles Atlas disciple and bodybuilder who pioneered the gym craze. The self-enhancement enterprise was bolstered by a burgeoning advertising industry that propelled a thriving consumer culture. Commercials transformed luxuries into necessities. Cars and refrigerators were no longer indulgences but essentials. Household gadgets were invented one after another: pop-up toasters, spin dryers for clothes, electric razors, to name a few. In accord with the buying spree and the inward focus, many men and women were keen to spend money on health treatments. The newfangled products were not seen as extravagances but as vital to well-being. Michael Pettit, in his thesis *Becoming Glandular*, called endocrinology in the 1920s the "technology of the self."

Steinach had not set out to devise a blockbuster youth-enhancing technique. His initial goals were more modest, more academic. As he told it, all he wanted to do was study the biology of sex glands in rats and perhaps shed light on human physiology.

Science advances by a combination of curiosity and skepticism. The best researchers do not read studies solely to pick up new information. They consider data and question it. Good scientists can't let go, particularly when they spot holes. They've got to hunt for the truth.

That was Steinach. One study both intrigued and irked him. In 1892, when he was a fledgling researcher, long before he was making headlines for his vasectomy cure, he happened upon a study of frog sex. The article described how male frogs stick to females like Crazy Glue and cannot let go until they ejaculate. The author posited a chain of events that started with a hormone gland and culminated with sticky feet. He wrote that when the male frog approaches the female, a fluid-filled organ tucked near the prostate and testes balloons, hitting against nerves that emit a signal, like an electrical signal, that travels upward and zaps the brain. The brain, in response, sends a signal along other nerves to the paws and increases their stickiness, so when the amphibians mate, they are stuck. When semen is released, the sperm-holding gland shrinks like a balloon releasing air, thereby easing the pressure on the nerves, which via the same pathway reduces the adhesiveness. In short, sex drive was fired up by a swollen organ pushing a nerve.

Steinach had his doubts. "But it seemed to me at least doubtful that a phenomenon so important to life as the instinct of propagation should be dependent on a local, variable factor as the filling and resultant distention of the seminal vesicles," he wrote. Seminal vesicles, tiny tubes tucked between the prostate and bladder, release the fluid that gives semen its sticky consistency. Nowadays we know that it was not just the analysis of the frog's sticky paws that was wrong; the whole account of frog sex was wrong. Frogs hug but they aren't stuck together. The male frog grabs the female tightly (it's called amplexus) and holds on until she vibrates and releases eggs that can be fertilized in the water by his sperm.

Steinach would go on to do a series of experiments, first disproving the nerve theory. Just as Starling showed that it is hormones, not nerves, that control the pancreas, Steinach would show that it is hormones, not nerves, that control sex drive.

In order to test the vesicle–nerve theory, Steinach removed the semen-secreting gland from four rats. If sex drive (in this case, measured by the urge to cling to a female) was controlled by nerves, the gland-free rats would have no urge whatsoever. But his gland-free males lusted after female rats. Steinach was excited. "What I actually saw almost bordered on the incredible," he wrote. "After the customary dallying the operated male rats repeatedly mounted females, who strenuously defended themselves. This sexual battle subsided to a certain extent after two days, but even in the later hours of the evening it was noticeable that the sexual excitability of the operated males . . . remained unweakened." He published his findings in 1894 in a German scientific journal, calling his paper "Untersuchungen zur vergleichenden Physiologie der männlichen Geschlechtsorgane" ("Investigations in the Comparative Physiology of the Male Sex Organs, Particularly of the Accessory Sex Glands"). He had proven an esteemed researcher wrong, but that begged a vital question: What drives the sex urge? Could it be hormones?

Steinach posited that scientists should hunt for hormone signals rather than tracing nerves. His thinking was similar to that of Cushing, who studied the pituitary glands of fat people, and to that of Berman, who studied the endocrine glands of the criminal. He believed that sex drive was under the influence of a hormone flowing through the blood; it was not managed by nerves in a tightly wound connection, like a mechanical toy. Before the beginning of the twentieth century, researchers had only a hazy notion that the causes of our urges were hidden inside tiny glands in the body, Steinach noted. Or, as he put it, "at first it was generally assumed that the whole complicated phenomenon was purely a nervous one, and that the only function of the gonads was to stimulate peripheral nerve endings."

Steinach saw more in the glands. He believed they had more power than nerves. He didn't dismiss the idea that nerves had something to do with sexual urges and with puberty, but he did not believe they were the key factor. He had so many questions. Could these internal secretions explain what made men men and women women? "Everyone knows even without books," he wrote, "that men are generally hardier, more energetic, and more enterprising than women, and that women show a greater inclination for tenderness and devotion, and a love of security together with a practical aptitude for domestic problems." Was he saying that a woman's ovarian hormones made her more inclined to stay home and nurture her man?

Following in the tracks of Arnold Berthold, the pioneering hormone researcher who performed the rooster testicle experiments in 1848, Steinach removed testicles from rats and watched the rats wither. Then, as Berthold had reimplanted testicles in the bellies of his birds, Steinach inserted the testicles into the rats' bellies. Voilà! The rats' energy increased and their sex drive surged. As had been shown more than a half century before, testicles work no matter where they dangle. More data that knocked down the nerve theory of bodily functions and promoted an endocrine one.

But what really piqued Steinach's curiosity—what really drove him to his rat cages—was the question of emotions and sex. Could the brain, could mood, sway hormones? In 1910, he devised an experiment to test whether male rats learned from other male rats to desire females, whether sexual desire emerged from something the females emitted, or whether it was inborn and caused by male hormones.

He placed ten male rats in cages, six alone and the others in a group of four. He kept all of them separated from females. When the rats were four months old, a female rat in heat was put into each cage. "All the male rats at once showed a distinct impulsion to these females, and immediately a vehement erotic playfulness took place, with the normal virility asserting itself in a pugnacious attitude towards strange male rats introduced together with the females." Simply put: the male rats competed with each other to get close to the females.

The males were then separated from the females and reintroduced to females in heat once a month. After eight months of separation and limited visiting rights, Steinach reported that the rats lost their sex drive. Even though they had the balls to be with women, lack of contact with the opposite sex dulled normal male desire. That proved, he contended, the ties between hormones and the brain: brain stimulation is needed to maintain hormone levels. He didn't monitor male-to-male attraction, nor did he care about female libido. His focus was the heterosexual male rat.

Next, he tried reversing the experiment. He put a wire mesh barrier in each cage and introduced a female, so potential mates could sniff each other but not have sex. Within weeks, the male rats returned to their lustful selves. "Without hesitation they immediately began the kind of chase which we had learned to recognize as sexual. This re-eroticization also showed itself in other signs of renewed virility, such as intolerance, aggressiveness, and jealousy towards rivals, who were at once violently assaulted."

When Steinach dissected the rats, he found that those who had been kept apart from females displayed shrunken seminal vesicles and shrunken prostates. Those who had played with the opposite sex had large ones. Steinach believed that this confirmed, once again, the psyche's powerful influence over sex drive, and that it disproved the nerve theory. (Steinach would forget about the power of the psyche a few years later, when he claimed that his vasectomy cure had everything to do with chemicals and nothing to do with the power of suggestion.)

For Steinach, each experiment solved one riddle but revealed another. After monitoring rodent foreplay, he mused about the sex-specificity of gonads. In other words, did the ovary or testicle act as a switch that activated an inborn femaleness or maleness? Were you born, say, a boy, and your testicles kicked the machine into action at puberty? If so, could you give an ovary to a castrated male (mouse, dog, or human) and turn this prepubescent youth into a full-grown man?

Steinach castrated two male guinea pigs and gave them ovaries.

He spayed the females and gave them testicles. Then he monitored their behavior and appearance. The males with ovaries but no testes grew large nipples, had smoother hair and a maternal instinct "with the same care, devotion, and patience as natural for a normal female." He spayed two females and implanted one with testicles. It grew a large clitoris, coarse hair, and "behaves exactly like a normal male," he reported, "scents at once which of the females is in heat, and immediately begins a vigorous courtship, which it pursues persistently, making repeated attempts at sexual contact. . . . the erotization [sic] of the brain has developed in a definitely male direction." In other words, the gonads, he decided, contained the essence of masculinity and femininity. He published his findings in 1912 in an article called "Arbitrary Transformation of Male Mammals into Animals with Pronounced Female Sex Characters and Feminine Psyche."

Steinach believed that he had found the clue to homosexuality: an abnormally high level of female hormones and not, as the thinking then was, bad parenting. No one, he declared, is pure woman or pure man.* He imagined that at some time in the fetal stage, babies are without a sex (the term gender wasn't used yet for living things) and had the potential to go either way; it depended on which hormone dominated and suppressed the other. If scientists could tap into that very early stage, they might be able to control the sex of the baby—but, he added, not the sexual orientation. "The most important decision in the life of a creature, the decision whether it has to go through life as man or woman," he wrote, "no longer

---

* Alfred Kinsey, a few decades later, believed in a spectrum, where everyone is ranked on the Kinsey scale of heterosexuality from 0 to 6, with few folks at either extreme. If you go to the Kinsey Center, you can buy a T-shirt with your own self-selected ranking.

appears as a matter of chance." Karl Kraus, an Austrian satirist, said he hoped Steinach would turn "suffragettes" into maternal women.

Steinach said his findings explained why some babies were born with ambiguous genitalia (called hermaphrodites then)—because their bodies did not suppress one sex. And he claimed that he could "cure" homosexuals by removing their testicles and replacing them with testicles from a heterosexual man. He reported that within the interstitial cells of homosexuals (the cells lining the sperm duct), he had found large cells not normally seen in men. He dubbed them F-cells, and said they looked like ovarian cells. Steinach suspected that these F-cells secreted female hormones. A few Dutch doctors concurred and added that Steinach's discovery would explain not only full-blown homosexuality but the "deviant" behavior in heterosexual men that occurred in all-male situations, such as prisons or boys' boarding schools. They called this pseudo-homosexual behavior.

So, what did all this have to do with vasectomies? Steinach combined evidence and assumptions to formulate his vasectomy-boosts-intellect-and-libido theory. He believed his research showed that gonads are intertwined with the psyche and that the more male hormone one has (testosterone wasn't isolated or named yet), the more one behaves like a lusting, aggressive man. His assumption was that if he destroyed one tissue, the adjacent tissue would overcompensate. When he blocked the sperm duct, for instance, he said that the hormone-secreting cells near it multiplied. Now scientists know he was wrong. Cells aren't weeds that overgrow just because the flowers nearby are picked.

In the late 1920s, Steinach tested his theory on geriatric rodents. They were two years old. "An old rat often presents a pitiable spectacle," he wrote. It hangs its head, sleeps most of the time, and totters apathetically when it meets a female. One month after being given a vasectomy, the rats "reawakened." They "became lively, inquisitive, and attentive towards what was going on about them," he wrote. "When rutting females were brought to them, they rose from their nests immediately to pursue, sniff and mount them. Reactivation was therefore achieved both physically and psychically."

On November 1, 1918, Steinach's friend Dr. Robert Lichtenstern performed the first vasectomy for the sole purpose of revitalization. His patient was Anton W., a forty-three-year-old carriage-driver. The patient showed up tired and scrawny, reporting that he had trouble breathing and could barely work. The operation was done under a local anesthetic. Lichtenstern sliced open the scrotum, snipped the vas deferens (the tube that carries sperm from scrotum to urethra), and tied the loose ends shut. In essence, he transformed a transit route into a cul-de-sac. (A vasectomy today is pretty much the same procedure, but with a smaller incision and without promises of rejuvenation or enhanced sexual prowess.) A year and a half later, Anton W. was a new man, or rather a younger-acting middle-aged man. His doctor reported that his skin had regained its smoothness, he had an upright posture, and he was able to work with renewed vigor.

Soon, doctors in Europe and the U.S. were writing to Steinach reporting spectacular results. Eighty-year-old men claimed they had regained their vim and vigor, along with their memories and business acumen. A New York City surgeon who Steinached an eighty-three-year-old stockbroker reported an "extraordinary improvement in the general state of his health." Before the procedure, he was a "tottering old man" and could barely work; soon afterward he was a thriving businessman who no longer had problems with urination, and his vision improved.

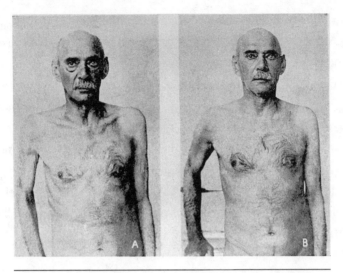

A 72-year-old man before and after the Steinach operation. *Courtesy of the Wellcome Library, London.*

Journalists loved the story. "Gland Treatment Spreads in America," declared the *New York Times* in 1923. "New Ponce de Leon Coming," was the *Baltimore Sun* headline announcing Steinach's American lecture tour. However, he never made it to the U.S. Claiming that he hated publicity, Steinach refused to give interviews. He blamed the American press for distorting the facts, insisting that he wasn't making everyone young forever. But he wasn't humble either. "I have shaken up mankind," he said.

Skeptical doctors were worried that Steinaching and other quack therapies would give the medical field a bad name and drive away the next generation of smart young potential doctors. "We have been Voronoffed and Steinached—Brinkleyed, too," wrote Dr. Van Buren Thorne in the *New York Times* in 1922. Dr. Morris Fishbein, the editor of the *Journal of the American Medical Association*, called Steinaching "hocus pocus." Other physicians ascribed the testimonials to the placebo effect. Steinach responded that studies disproved that assertion; several doctors, he said, had given vasectomies to unwitting patients just to test their effectiveness. Men had come in for a hernia operation, say, or some other operation in that region, perhaps to remove a cyst, and unbeknownst to themselves had received a vasectomy too. Months after the procedure, when the patients were asked if they were feeling smarter or more sexual or younger, the surgeons claimed that everyone said yes. (It should be noted that the secret vasectomies were done long before the days of mandated informed consent, and that now patients sign forms spelling out precisely what is going to happen in the operating room.)

But did this really prove the effectiveness of the Steinach procedure? The volunteers may not have known they had had a vasectomy, but they knew something had been done to improve their health. When their doctor asked whether they were feeling better, they may have preferred to answer positively. More importantly, the reported successes were all based on testimony that was far from the gold standard today: the double-blind randomized controlled trial. Steinach did not, as would be expected today, divide men into groups and give half the real operation and half a sham one. He did not ensure that neither patient nor evaluator knew who got which operation; that's the double-blind part. Such trials, now de rigueur, were not routine until the mid-twentieth century. Steinach did what was appropriate for his time.

The spate of operations and the positive publicity spurred popularity for his methods—except when the publicity didn't go as planned. Alfred Wilson, a Steinached septuagenarian Englishman, was delighted with his results. He had paid £700 for the vasectomy and wanted to share his rejuvenation with the public. He rented

London's Royal Albert Hall—which had hosted German composer Richard Wagner in 1877 and would go on to host the Beatles in 1963. His plan was to stand there and look robust and take questions from the audience. His one-man manly show, "How I Was Made Twenty Years Younger," was sold out for a performance to be held on May 12, 1921. But the night before the highly anticipated evening, Wilson dropped dead of a heart attack. That made for a great story for the tabloids. Steinach maintained that the operation had nothing to do with Wilson's death.

For a time, the popularity of the procedure continued. Norman Haire, an Australian gynecologist practicing in London, wrote a book about Steinaching, pointing to more than two dozen successful Steinachs that he had performed. (Odd that a man would go to a gynecologist for a vasectomy.) One of his patients was a fifty-seven-year-old doctor who said his erections became vigorous and the operation repaired "some disharmony in the relations with the wife, who is a great many years younger." In 1929, at a meeting of the International Congress of the World League for Sexual Reform in London, Peter Schmidt, a German doctor, announced that he had performed 600 Steinachs, all with good results.

Today, we know that, though a few studies suggest that you may get a fleeting blip—the tiniest uptick—of testosterone after a vasectomy, for the most part it does not have an impact on hormone levels. In other words, after a vasectomy, nothing changes except that you are no longer releasing sperm. Yet what eventually punctured the phenomenal success of the Steinach operation was not the faults of the surgery or its uselessness as a libido-booster, but rather the isolation of hormones. There would be easier options—a drug instead of surgery.

Steinach missed the mark when he explained how his operation worked—by causing an overgrowth of interstitial cells. It didn't. But he got a lot right. He established that the interstitial cells are the main source of sex hormones in males. He pioneered ideas of sexual behavior as a complex interplay between gonads and the brain that included nerve input as well. And, though it may not fit into the getting-a-lot-right bucket, he helped to promote the lucrative business of sex hormones. He created the market for hormone-based rejuvenation drugs.

Despite all the craziness, there was plenty of serious science. Investigators were tapping into the newest fields of laboratory research, combining biology and chemistry. Many of the discoveries made headlines—the isolation of estrogen, progesterone, and then testosterone,* but among them was one often overlooked hormone finding. The initial research began in Germany in the late 1920s and culminated nearly a decade later in Baltimore, where a brash young female medical student had the gall to think she could solve a medical mystery.

---

* Estrogen and progesterone were isolated in 1929, testosterone in 1931.

6.

Soul Mates in Sex Hormones

FOR NEARLY HALF A CENTURY, Dr. Georgeanna Seegar Jones shared a desk with her husband, Dr. Howard W. Jones, Jr. It was a partners' desk, a large antique mahogany table with drawers on opposite sides so that two people could work in a space built for one.

The Joneses were known for other things in addition to their lifelong devotion to each other. In 1965, they worked with Cambridge University's Robert Edwards and fertilized a human egg in a lab, something that had never been done before. Edwards created the world's first test-tube baby in 1978. Three years later, the Joneses performed America's first successful in vitro fertilization and helped launch the modern fertility business.

For all the test-tube baby hoopla, few folks realize that the Joneses launched the modern fertility business *after* they retired. And even fewer folks realize that Georgeanna Seegar Jones made an impact on the field of endocrinology long before babies were

Howard and Georgeanna Jones, 1958. *Courtesy of A. Aubrey Bodine.*

made in Petri dishes, and long before women began to make headway in this all-male field.

It all began on a fateful evening: February 29, 1932. A leap day. Georgeanna Seegar, a college senior, attended a lecture at Johns Hopkins Hospital because her father, an obstetrician–gynecologist, insisted. The pioneering neurosurgeon and endocrinologist Harvey Cushing was the speaker. The evening also sparked the romance between Seegar and Jones. Jones's father was also a doctor, and Seegar's father had delivered Jones in 1910. As toddlers, the two had played together on the hospital lawn while their fathers saw patients on weekends. For decades, Seegar would say that the Hopkins evening changed her life. Or as Jones loved to tell it, "I, of course, thought she was referring to the fact that the two of us had recently re-met, but not at all: she was referring to the fact that Cushing had so activated her by his description of abnormalities of internal secretion that she determined that endocrinology, then a relatively new subject, would be her specialty field of ObGyn."

Seegar, as her father hoped, enrolled in Johns Hopkins medical school the following year. The lovestruck Jones, a second-year student, sought out Seegar at her anatomy table. She shared a cadaver with Al Schwartz, a friend of Jones's since his undergraduate years at Amherst. That provided a good excuse for Jones to wander by and strike up a relationship. Dating, in Hopkins parlance, usually meant studying together in the library. For Seegar and Jones, it also meant examining pieces of ovaries under a microscope in the pathology lab. Finally Schwartz told Jones that it was about time he asked Seegar on a real date.

Jones mustered up his courage and invited Seegar to go horseback riding on Thanksgiving, as he had heard she liked to ride. But that morning there was a drenching rain. "Somehow or other I did not suggest another activity," Jones recalled some eighty years later. "Why I don't know, but I just didn't." About a month later, the two finally went on their first date. It was a brisk, sunny New Year's Day, 1933. They drove out to stables north of Baltimore.

"The women who went into medicine in that era, we referred to them as hen medics and that implied a certain affect, which was not exactly a feminine affect: flat shoes, plain dress," said Jones. "I refer to that as an academic look. You looked like you had little chance of getting married." Jones was one hundred years old, flipping through black-and-white photos of himself and Seegar in their younger days, and the details were fresh in his mind. "She was not like that," he added. "High-heeled shoes. She dressed well, groomed well."

When they were in their first years of medical school, Jones figured out a way to ensure that he and Seegar had dinner together at least once a week. He launched a club, inviting a dozen other students and the mandatory faculty advisor to make the whole thing legitimate. The point of the club—other than seeing his girl—was to discuss a recently published medical book: *Sex and Internal Secretions*, a hefty treasure trove of everything anyone would want to know about the burgeoning field of sex studies and sex hormones. It was edited by Edgar Allen, a professor at Washington University in St. Louis. Allen, along with Edward Doisy, had gained prominence for purifying estrogen in 1929. The book included chapters by an eclectic group of hormone researchers, including physiologists, biologists, and psychologists, as well as entomologists and ornithologists. It started with insect mating habits, veered into bird plumage, and climaxed with the physiology of human sexual abnormalities.* It was a serious and complicated medical tome that touched on the latest theories of sex, sexuality, and the physiology of puberty.

Jones called his gatherings the Sex Club. They convened every Friday at five o'clock at the Shop, a popular greasy spoon on the corner of Wolfe and Monument, a five-minute walk from campus. Over burgers and milkshakes, they discussed the book one chapter at a time. On the first meeting, students got their burgers and delved into the basic biology of sex differentiation. It was already known

---

* Interestingly, Kinsey's early work was with insects, specifically wasps. Perhaps there is a natural tendency to go from insect biology to human sex.

by then that all embryos start out the same, but something, perhaps some chemical trigger or something in the environment, such as the mother's diet, must occur to push an embryo to become female or male. The question that came up again and again in *Sex and Internal Secretions* was this: What controls maleness and femaleness, and what do those labels mean? Is it all to do with chromosomes? Or hormones? Or something else?

The concepts were new, muddled yet titillating. Students read about "conditioned" and "unconditioned" sex. The idea was that female hormones prompt the female embryo to develop female anatomy; the same was true of males. A body part without any hormonal stimulus was "unconditioned." An example of an unconditioned sex characteristic was the mammary glands of a male.

The students read that an embryo could switch sexes, or become a bit of both, if something interfered in the process of sex differentiation. For instance, in one experiment, scientists had created an intersex calf (called a hermaphrodite calf then) by injecting blood from a male calf into a female embryo. "Every zygote," wrote Frank Lillie, the dean of biological sciences at the University of Chicago who authored the first chapter, "is thus potentially hermaphroditic in the sense that it is capable of giving rise to the characters of either sex or subject to conditions of determiners, to characters of both sexes, i.e. to individuals that are actual gynandromorphs or intersexes."

The book captivated Jones and Seegar. Indeed, when they got through their first reading—all 910 pages—they started it again. And Seegar did more: when she wasn't in class or at the Sex Club, or preparing for their weekly meetings, she volunteered in the lab of George Otto Gey (pronounced Guy). Her job was to run the pregnancy tests. What started as a lab technician's task led to her momentous discovery, because Seegar wasn't just going through the motions; she was thinking about each step along the way, pondering the science of it all.

The pregnancy test then was called the A–Z test, after its inventors, Selmar Aschheim and Bernhard Zondek, German doctors. A urine sample from a possibly pregnant woman was injected into a mouse, and after about 100 hours, the mouse's ovaries were examined. If they were swollen and speckled red, the woman was pregnant. If there was no change, she wasn't. It was cumbersome by today's pee-on-a-stick standards, but back then it was considered speedy and simple. Before the Aschheim–Zondek hormone test, a woman had to wait until she missed two or three periods and doctors heard a heartbeat—months into pregnancy.

In the early 1930s and for the next few decades, the A–Z test was *the* pregnancy test. Eventually, mice were replaced with rabbits which were replaced with frogs, which was a good thing as frogs didn't have to be killed since they lay eggs. (This is the genesis of the nonsensical 1950s euphemism for pregnancy, "the rabbit died"; it made no sense because the rabbit was killed either way.)*

The doctors had christened their pregnancy test *hypophysen-vorderlappenreaktion*, a tongue-twister even by German standards. It means "anterior pituitary reaction," in reference to the gland that they thought released the vital hormone. The name didn't catch on, for reasons other than its ten laborious syllables. It left scientists scratching their heads. Why the pituitary? Where was the proof? Spiked pee came from the brain—really?

Ever since Harvey Cushing fixated on the pituitary, scientists had assumed it was responsible for anything hormonal. Indeed,

---

* In the 1950s, my mother went in for a pregnancy test and, in addition to the doctor's bill, received a bill in the mail for three dollars to pay for the rabbit. She wasn't pregnant that time.

Aschheim and Zondek had injected chunks of pituitary into rodents and found that the little creatures ovulated. The ovaries seemed to respond just as they did when injected with the urine of pregnant women. It made sense to Aschheim and Zondek that the chemical in the pituitary was the same chemical as in the urine, but they had never isolated the substance from the pituitary. It was good, but not conclusive, evidence.

It was good enough for most doctors, particularly as Aschheim and Zondek were well-respected in the field, but some experts were skeptical. Earl Engle, a Stanford University researcher, found that, on close inspection, urine from pregnant women and pituitary injections caused eggs to respond differently. After a shot of pregnant woman's urine, rodent follicles burst out of the ovary, blood vessels swarmed to the maturing eggs, and the corpus luteum, the egg's nutrient supply, grew. After a shot of pituitary, the follicles were released but nothing else happened: no blood congestion, no corpus luteum growth. (The corpus luteum, a yellow blob within each egg, is crucial to pregnancy. It releases hormones that convert immature eggs into ones that can be released and fertilized. It also releases hormones that sustain the pregnancy, pumping estrogen and progesterone that signal the body to cushion the lining of the womb.)

Another study, a tiny one that included merely a dozen mice, showed that pieces of placenta triggered precisely the same thing that the urine of pregnant women did. When a mushed mouse placenta was injected into young mice, scientists witnessed swollen follicles (the eggs), congestion (blood flowing to the area), and the development of corpus luteum. That suggested that whatever is in the urine could be in the placenta as well.

In 1933, Georgeanna Seegar entered the fray of scientists

hunting to isolate the hormone from humans. She was one young, unknown female student among a gaggle of esteemed male professors. The lab where she ran the A–Z tests was home to a brand new apparatus: a roller tube machine, as it was known, which kept cells alive so that doctors could study them. From today's standpoint, it seems awfully ordinary, but back then it was a colossal achievement. Gey had built the machine, creating something researchers had long yearned for, a way to grow human cells outside the body. The device would enable them to learn about live biological processes as they happened, rather than staring at a dead chunk of tissue. It also provided a way to test new treatments. Investigators had, previously, tried to keep cells alive in Petri dishes, but even with the best nutrients they'd wither away. In Gey's machine the cells received a constant supply of fresh nourishment—like standing under a shower rather than soaking in a stale bath.

A do-it-yourself kind of person, Gey had collected scraps of glass and metal from Jake Shapiro's junkyard in Baltimore, using the glass to blow his own test tubes which he slid into slots in a homemade metal drum. The tubes, containing cells and their nutrient substrate, rotated ever so slowly, about one turn an hour. The movement pushed the cells to the glass wall as their food washed over them. Pulses of carbon dioxide (which maintained the right level of acidity) wafted into the drum. It was this same roller tube machine that would, in 1951, be used by Gey to create the immortal HeLa cell line, retrieved from Henrietta Lacks's cervical tumor and used for years for all kinds of medical studies.

Gey was a fast talker and fast thinker who juggled a lot of projects at the same time, always trying to concoct new ways to do things. His wife, Margaret, who had trained as a nurse, was his medical technician. She was the workhorse, doing the nitty-gritty of her husband's lab work, making sure his ideas were executed carefully. The Geys, Seegar, and Jones (who was often visiting his girlfriend) ate brown bag lunches together, which provided the students an opportunity to eavesdrop on the latest scientific debates. And so it was that over sandwiches at the Hopkins laboratory Seegar heard about the placenta–pituitary results. She asked if she could use Gey's new cell culture machine to test the placenta hormone theory.

Gey had no problem with Seegar using the machine, but couldn't imagine she'd be able to accomplish the feat. Obtaining a placenta was not a simple task. Gey explained that she couldn't use one that had traveled through the birth canal because you wouldn't know if the placenta had made the hormone or if it had picked the hormone up on the way out. A placenta from a cesarean section would be acceptable, but in those days C-sections were rare—only about 2 percent of births. Seegar wasn't dissuaded.

As luck would have it, Seegar's friend Louis Hellman found an unadulterated placenta for her. He was a resident at Johns Hopkins, but was spending a few weeks at a Harvard lab, where he came across a rare, hormone-spewing growth that had been removed from a womb. The woman had thought she was pregnant because the pregnancy test had come back positive, but she wasn't. Instead of a sperm fertilizing an egg, just bits of sperm had entered the egg, which happens every now and then. It triggered a tumor-like growth and the formation of a placenta. Knowing what Seegar needed, Hellman got permission from Arthur Hertig, the lab's director, and hopped a train from Cambridge to Baltimore with the placenta in a jar.

Seegar was thrilled. She crushed the placenta and put it into Gey's roller tube machine so that the cells would proliferate. Then came the real test: would the placental cells change the ovaries of young female rats? That would be a sign that the placenta contained the baby-making hormone. She injected the sample and—presto! The placenta triggered the same changes as the urine of pregnant women had: congestion, follicular growth, and corpus luteum breakdown. It was a small study—one placenta—but it was scrupulously done. It was considered a solid piece of evidence supporting the notion that it was the placenta that harbored the pregnancy hormone, not the pituitary.

Gey told Seegar to publish the findings right away, while they hunted for more placentas with which to confirm the results. Gey suggested writing a letter in the journal *Science*, which would be speedier than submitting a scientific article. That way, they could plant their victory flag as solvers of the debate. The only glitch was that, as Gey knew, a paper authored by a woman would have a slim chance of being accepted by a respected journal. He told Seegar to sign it using her first initial and middle name instead of her first name: G. Emory Seegar.

The letter was published in *Science* on September 30, 1938. Gey put his name first as the lead author—customary, because he was the most senior person in the laboratory. Seegar felt slighted but didn't complain. Margaret Gey, who shared the bulk of the laboratory work with Seegar, didn't even get a mention. The paper had three authors: Gey, Seegar, and Hellman, the student who had carried the placenta from Harvard to Johns Hopkins.

The following year, Bernhard Zondek, who still believed in the pituitary origins of the pregnancy hormone, was invited to Johns Hopkins to lecture. A dinner was planned at the exclusive Maryland Club. Seegar, because of her role in the study, was the only student invited. Protocol required that anyone who was not a Maryland Club member had to be approved before attending a meal. The club informed Emil Novak, the gynecologist who organized the event, that Seegar was not welcome as the Maryland was a men-only club. Novak, infuriated, retorted that "Georgeanna is the most important guest, in view of her work, and if she can't come, we will take the luncheon elsewhere." The club reluctantly relented.

In 1942, the team published a larger study in the *Bulletin of the Johns Hopkins Hospital*. They also renamed the pregnancy hormone. Aschheim and Zondek had called it prolan, from *proles*, the Latin word for offspring. Seegar called it chorionic gonadotropin. This described the hormone for what it truly was: a substance that nourishes the gonads and comes from the chorion, part of the placenta. The paper contained the results of seven placentas: two ectopic pregnancies (when the embryo grows outside the womb, in the fallopian tubes, and has to be removed), two full-term pregnancies delivered by C-section, and three hydatidiform moles, the tumor-like growths that had triggered the formation of the original placenta studied. Seegar presented the findings in New Orleans at a meeting of the American Physiological Society on March 15, 1945. Her label, chorionic gonadotropin, stuck; she later added the word "human." The hormone is now called hCG for short. It's often given to women undergoing fertility treatment—the in vitro fertilization procedure that the Joneses pioneered—to boost the chances of success.

Seegar not only solved a medical riddle; she named the pregnancy hormone and became the first women ever to dine at the Maryland Club. And two of those things she did before she graduated from medical school.

In those days, medical students were not allowed to be married.

Howard Jones and Georgeanna Seegar wed the day they finished their training in 1940 in a simple church ceremony. A few years into their marriage, Jones was drafted into World War II. Seegar applied to join him at the front. No matter that they had a newborn and a two-year-old; relatives and a nanny could, she said, take care of the kids. The medical examiner thought differently and turned her down.

While Jones was gone, they figured out a secret way for Seegar to know where he was stationed, as he wasn't allowed to write his location in letters. They each bought the same European map before he left. When Jones wrote a letter, he would place the stationery over the map, with the edges always in the same place, and poke a tiny hole in the paper over his current bivouac. When Seegar received the letter, she'd line it up over her version of the map and search for the pinhole.

After the war, they had one more child, and returned to their medical careers, working together at the partners' desk. From then on, they would be referred to by their patients and colleagues as Dr. Howard and Dr. Georgeanna.

Dr. Georgeanna had short-cropped hair and wore conservative dresses, and had a precise and self-assured manner that exuded authority among the sea of men. Her bedside manner with patients differed from the masculine norm. In the middle of the twentieth century, when doctors were trained to be distant, Dr. Georgeanna was soft-spoken and compassionate. One patient remembered her leaning over and whispering encouraging words while she was being wheeled into the operating room for a fertility procedure.

She was never swayed by popular opinion and was a stickler for data. In the 1950s, she advised her staff never to prescribe DES, a drug made from synthetic estrogen designed to prevent miscarriages that was widely used at Harvard and other major institutions. She had scrutinized the evidence and wasn't convinced. In 1971, twenty years after its launch, DES's toxic side effects were revealed: it triggered vaginal cancer, uterine deformities, and infertility in women who had been exposed to it before birth.

In her final years she suffered from Alzheimer's disease. Dr. Howard stopped practicing medicine at the same time. "It just wasn't fun anymore without her," he said. But he continued to go to the office to read and attend conferences, and brought Dr. Georgeanna with him as long as he could. Nancy Garcia, their administrative assistant, kept a curling iron in the office to do her hair. "Dr. Georgeanna would walk into Dr. Howard's office and he would say, 'You look beautiful, Georgeanna,'" Garcia recalled. "That's all she wanted to hear." Georgeanna Seegar Jones died on March 26, 2005, at the age of ninety-five.

The Joneses' lives encompassed the history of reproductive endocrinology. The two of them were often front and center, playing a pivotal role. Their protocols and research would provide new avenues for understanding the hormones of sexual development, but would—years after they retired—become the focus of highly contentious debates.

7.

Making Gender

WHEN CATHLEEN SULLIVAN gave birth in the summer of 1956 in New Jersey's West Hudson Hospital, the doctor pulled the baby out with forceps, looked between the newborn's legs, and said . . . nothing. This was Ms. Sullivan's first child and she didn't know what to expect from her family doctor right after the baby came out, but imagined some kind of "it's-a-girl!/it's-a-boy!" pronouncement. He didn't utter a word.

The doctor was flummoxed. The baby's genitals looked neither female nor male, but something in between. And he didn't know how to reveal these doubts to his patient. It's hard enough for a doctor to express uncertainty in any situation, but when it comes to something as basic as a baby's sex, it was all the more humiliating. How could he *not* know? How could he admit that he didn't know? So, as Ms. Sullivan emerged from the fog of childbirth drugs, he sedated her again, buying time so he could get some advice to help him decide whether the baby should be a he or a she. Arthur

Sullivan, Ms. Sullivan's husband, never talked about those first few days to friends or, later, to their children, so no one knows what he was told, if anything. The doctor certainly couldn't sedate him too.

Three days later, Ms. Sullivan was handed her baby. A boy, the doctor announced, albeit one with a serious deformity. Genital surgery might help, he told the new parents, but they wouldn't be able to do anything for years. Mother and son went home. They never heard from that doctor again. Ms. Sullivan tried to get in touch with him, but he did not return her calls.

The child was named Brian Arthur Sullivan. By every measure except genital appearance, young Brian was like any other baby. He reached every milestone on time, if not sooner. And yet, from the very first coos and wriggles, his parents worried. His penis was tiny, too tiny even for a baby. It didn't have a foreskin. Ms. Sullivan had heard stories from other new moms about dodging their little boy's projectile urine while changing them. But Brian leaked from the bottom of his penis, more like a girl. For now, his secret was tucked in his diapers. But what would happen when he got older? Would classmates make fun of him? Would he have to pee sitting down? Would other parents find out? Would he ever fit in?

When a baby doesn't conform to the standards of normalcy, parent can't help but wonder if it is somehow their fault. Did they harbor some genetic mutation that was passed down to their son? Did they do something wrong during pregnancy? Ms. Sullivan was on bed rest during her fifth month because of bleeding. Was that a missed signal that something had gone terribly awry? The Sullivans were frightened and, well, embarrassed. They had no guidance, no help.

If their baby had had asthma or diabetes, they could have reached out to friends for sympathy and advice. But in the

1950s—when even adoption and infertility were taboo subjects—how could they bring up atypical genitalia? Without any kind of support network, the new parents were left to navigate this thorny issue alone. They did their best to protect young Brian and to keep things as normal as possible, but their own fears cast a shadow over the way they treated him, the way they looked at him. The Sullivan child would recall years later that it seemed as if they were always angry at him, as if they were monitoring his every move, as if nothing he did made them proud like parents are supposed to be.

If Brian didn't have all the usual boy parts—as his small penis and unusually shaped scrotum (it was empty and completely open in the middle) seemed to suggest—did he have a deeper problem, they wondered?

Ms. Sullivan's sister, one of their few confidants, suggested they get advice from a specialist and found a doctor at Columbia University, not far from the Sullivans' home in Kearney, New Jersey. Doctors at Columbia, and also at Harvard, Johns Hopkins, the University of Pennsylvania, and other leading institutions, were studying children like Brian, trying to figure out if an unusual mix of hormones had anything to do with their abnormal genitalia. An overload of female hormones? A dearth of male ones? Maybe the Columbia doctors could tell Brian's parents what had gone wrong. Maybe they could even prescribe something to fix things. Maybe they would tell them that kids like Brian were late bloomers and by the time he started school, he—or more precisely, his penis—would look just like the other boys.

The Sullivans saw the specialist when Brian was three months old. The doctor examined the baby and told them to come back in nine months. He didn't explain anything to them, never say-

ing whether he was worried or suspected a disease. The Sullivans
didn't ask.

Brian celebrated his first birthday. He chattered, scampered,
played with trucks and blocks. Then there was another baby in the
house. Mark was born when Brian was ten months old. Though
the Sullivans continued to fret over their eldest, what with all the
sleepless nights and demands that go along with a newborn they
didn't go back to Columbia when they were supposed to. Finally, in
the last week of January 1958, when Brian was seventeen months
old, his parents bundled him in his blue snowsuit and returned to
Columbia Presbyterian Hospital. This time, the doctor suggested a
thorough examination. Sometimes, he explained, when things do
not appear typical on the outside, it can signal something going on
in the inside. He wanted to perform exploratory surgery, which
meant cutting open the baby's belly to see the reproductive organs.
If anything needed to be fixed, he could do it right away and report
back to the parents afterward.

Few patients in the 1950s expected handholding, a sympathetic voice, or a detailed explanation from a doctor. And few folks
did what patients—or parents of patients—do today: grapple with
doctors about treatment options to figure out which one is best. For
the most part, doctors were considered to know it all. And, chances
were, the doctors considered themselves wise men, too. (There
weren't many women.) They were the ones who had spent years in
medical school. Neither they nor their patients saw any reason for
a crash course in biology during a thirty-minute meeting. Medical
experts didn't need anyone's naïve input.

These were the days before patients would advocate for a so-
called Patients' Bill of Rights, which was conceived in the 1970s
to spell out what to expect at a hospital visit. This was before

informed consent, the contract that documents that the doctor has told you everything about your diagnosis, including potentially nasty side effects; before "health partner" and "patient advocate" were part of the health care lexicon. If a doctor thought a patient couldn't handle, say, a cancer diagnosis, then he might just not mention it. Medicine was a paternalistic enterprise, its dominance enhanced by a vast armamentarium of drugs and therapies, many available for the first time in the 1950s. Doctors had the wherewithal and the power to do what they and they alone deemed appropriate, often without approval from patients or third-party payers.

So when the doctor told the Sullivans to leave Brian with him for a few weeks—or maybe even longer—and let him handle all the decisions, it wasn't considered callous or overly authoritarian. It was just the way things were. Ms. Sullivan didn't confer with the doctor but she visited the ward every day, driving to New York City from her home in New Jersey and sneaking in a pacifier, which for some odd reason was against hospital rules. About three weeks later, the doctor told the Sullivans that he had figured out the problem. He had found a uterus, vagina, and ovaries inside Brian's abdomen. Brian's penis wasn't a penis after all; it was a large clitoris. Brian was not a boy. Brian was a girl.

Because the clitoris was too big, the doctor said, he amputated it, to ensure that there would be no strange looks in the school bathroom or at sleepovers, no risk of kids giggling behind their now-daughter's back about her odd-looking genitals. The Sullivans were told that, thanks to the operation, their child looked like a normal girl stark naked.

And then the doctor matter-of-factly explained how they should start treating their new daughter like a daughter. First, a name change. How about Bonnie? That had a nice ring to it and sounded like a female version of Brian. So Brian Sullivan became Bonnie Sullivan. On Presbyterian Hospital letterhead, the doctor told Ms. Sullivan to sign, as the legal guardian of her child, this note:

THIS IS TO CERTIFY THAT MY NAME PREVIOUSLY WAS
Brian Arthur Sullivan
AND IS NOW *Bonnie Grace Sullivan*

The doctor listed the steps required to complete the transformation. Their new little girl needed a complete makeover: a feminine wardrobe (pink dresses, no pants), longer hair (she would eventually have a stylish bob), girly toys (dolls, no trucks). The doctor also recommended that the Sullivans move to another neighborhood in order to solidify their daughter's new identity. They should find a place where nobody had heard about Brian, where Bonnie could start life again with a clean slate, where everyone would treat her like a fully-fledged girl, where no one knew about the switch. He reassured the Sullivans that if they abided by a set of guidelines that he would elucidate, their young daughter would, as the song goes, feel like a natural woman.

The experts instructed the Sullivans to comb through the house and get rid of baby pictures, home movies, birthday cards, or anything else that documented Brian's previous existence.

The Sullivans tried their best to delete "Brian," but they didn't leave their New Jersey town for a few years, not until Bonnie started first grade. They had aimed to move by kindergarten, but with work and life and whatnot and then another child, it wasn't easy to pick up and move to another town with three little ones in tow. (Another child, a daughter, was born six years after Brian/Bonnie.) Ms. Sullivan confided to her next-door neighbor, who was sympathetic and bought Bonnie her first doll.

Had Bonnie been born fifty years earlier, she might have been one of the so-called freaks displayed in the circus, next to the Fat Bride or the Armless Wonder. It's hard to imagine the entertainment value, but turn-of-the-twentieth-century amusement park managers hired people born with atypical genitalia too. If Bonnie had been born fifty years later, the doctors might have discussed the options before launching into surgery. Perhaps her parents would have postponed the decision until Bonnie was a teenager and could have a say. Maybe her parents would have found the burgeoning advocacy groups that promote living happily without surgery.

But Brian/Bonnie was born in 1956, right at the cusp of a new era in endocrinology. Doctors had better insight into how testosterone and estrogen molded external genitalia and sexual development. They appreciated the chain of command among the glands: the adrenal hormones, for instance, were under the control of the pituitary gland, which in turn was under the control of another brain gland, the hypothalamus. This new understanding fueled the way specialists diagnosed and treated children born with ambiguous genitalia: a way that combined new hormone drugs, state-of-the-art psychological assessments, and new operations. Doctors were also more willing to operate than before, thanks, in part, to the recent discovery of antibiotics, which drastically reduced the risk of postoperative infection. "The last decade has witnessed a revolutionary change in the treatment of intersexuality," Howard W. Jones, Jr., told colleagues at an American Gynecological Society meeting in Colorado Springs in 1961. This was the same Jones who was the test-tube baby pioneer and gynecologic surgeon married to Georgeanna Seegar Jones, the pregnancy hormone discoverer. "It would be naïve to believe that all the problems of intersexuality have been solved, but within the last decade great strides have been made."

It had never occurred to the Sullivans that a baby could be born with anything other than genitalia marking it clearly a boy or clearly a girl. They had known all along that Brian/Bonnie didn't look like a typical newborn, and had worried about the child's masculinity. Yet they never thought he wasn't a boy. After raising him for a year and a half, Ms. Sullivan missed her son. She felt as if the baby she had birthed no longer existed. One moment she was the mother of a beautiful baby boy, then nearly eighteen months later, the doctors were telling her it was a case of mistaken identity.

But that's not what was written in the baby's medical chart. The first doctor wrote "Hermaphrodite."

Hermaphroditus, in Greek mythology, was a teenage deity seduced by a nymph. Though he rejected her advances, she wrapped her body around his and begged the gods to merge them forever. He was no longer a he; she was no longer a she. They were one. They were both and neither. Another version of the myth holds that Hermaphroditus inherited his father Hermes's brawn and his mother Aphrodite's beauty, so his name and his body were a union of both parents, representing the ideal human.

Whatever the origins of the name, doctors adopted the word "hermaphrodite" to describe children like Brian/Bonnie. The standard medical textbook was titled *Hermaphroditism* (Howard Jones was the coauthor of both editions). The term stuck until about the 1990s, when patients insisted they be called something without the circus-sideshow connotations. The label "DSD" has been introduced, standing for "disorder of sex development" or "difference of sex development" (as many people balk at the word "disorder");

many want to jettison "DSD" altogether and prefer the term "intersex," as it does not imply abnormality.

Today, ambiguous genitalia is reportedly as common as cystic fibrosis, a lung disease, but much less talked about. The statistics are rough, ranging from about one in 2,000 to one in 10,000, depending on which conditions are included. It's likely that if you attend a large university or work for a big company, you've come across someone whose life is significantly affected by intersexuality, but you may not know it.

In our very first weeks in the womb, we all look alike: a furiously multiplying orb of cells. Then the sphere stretches to a curved oblong, like a dinner roll turning into a croissant. At one end is the developing brain; at the other is something that looks like a vagina—a fold with a little knob at the edge. A flick of a switch—hormonal tugs this way and that—molds the unisex fetus into a boy or a girl. In a sense, we all begin as hermaphrodites.

In the early 1900s, the University of Chicago's Frank Lillie noticed that when blood vessels got mixed up during the gestation of male–female calf twins, the female was born lacking a uterus and ovaries. That got him thinking that something—a chemical perhaps—in the blood of the male fetus staunched the female's development. To test his theory, he injected blood from a male calf fetus into a female one. Lo and behold, the female calf was born intersex, with some female external genitalia but no reproductive organs inside. "Every zygote is thus potentially hermaphroditic in the sense that it is capable of giving rise to characters of either sex," Lillie wrote in the textbook that Georgeanna Seegar and Howard Jones devoured during their weekly Sex Club meetings. The zygote, he went on, can also give rise "to individuals that are actual gynandromorphs or intersexes."

Alfred Jost, an endocrinologist working at the Collège de France in Paris, pinpointed the precise hormone that kicks in at six weeks of gestation, by studying male rabbit fetuses. He named the chemical anti-Mullerian hormone. The Mullerian ducts—named for Johannes Peter Müller, the investigator who described them in 1830—grow into female organs. Anti-Mullerian hormone spurs the development of male organs, the scrotum and testes, while suppressing female ones, the ovaries and uterus.

Boys have anti-Mullerian hormone. Girls don't. In girls, there's nothing to fire up the male track and nothing to stop the female one. Because girls are made by the absence of a hormone, scientists have long considered femaleness the default path—which sounds like a consolation prize. According to Rebecca Jordan-Young's *Brain Storm*, recent evidence suggests that femaleness may not be merely a default after all; the ovary may have signals of its own. Still, the notion that females are created by a passive process persists among many scientists.

Of course, the system is more complicated than that. There are many genetic signals that must be activated, and hormones that must be released at precisely the right time in precisely the right dose. It's a wonder that any of us are born within the so-called conventional zone.

The umbrella term "intersex" includes many conditions. In Bonnie Sullivan's time, such children were labeled "true" or "pseudo-" hermaphrodites. Bonnie was in the "true" category because she had both testicular and ovarian tissue. Girls born with a condition called congenital adrenal hyperplasia, or CAH—which describes a block in the cortisol pathway, promoting too many androgens—were put into the "pseudo" group. The advent of synthetic cortisol in 1949 made it possible to replenish the insufficient cortisol and alleviate the androgen symptoms; for CAH children who also lacked aldo-sterone, an adrenal hormone that balances salt and water, synthetic cortisol was literally a lifesaver.

Nowadays, scientists have a much clearer understanding, some-times right down to the tiniest genetic hiccup. Some XY fetuses, for instance, do not respond to the testosterone secreted by their testes and are therefore born with female genitalia, though not a uterus or vagina. Other XY children lack an enzyme (5-alpha-reductase type 2) that transforms one form of testosterone to another, necessary for the formation of male genitalia. They look like girls at birth. But they do not lack the other form of this enzyme (5-alpha-reductase type 1) that kicks in at puberty, switching their appearance from feminine to masculine. Many of these individuals end up not fully masculinized.

Bonnie's parents weren't given a name or a scientific explana-tion for her disorder. They simply followed the doctors' orders. The doctors, in turn, followed guidelines established at Johns Hopkins Hospital in Baltimore, the premier center for the study and treat-ment of children born with ambiguous genitalia.

Hopkins not only pioneered hormone therapies, includ-ing using cortisol for congenital adrenal hyperplasia, but also established an interdisciplinary approach, recruiting top-notch psychiatrists, reproductive endocrinologists, plastic surgeons, urologists, and gynecologists. Georgeanna Seegar Jones, the pio-neering reproductive endocrinologist, was involved in the hor-mone aspects of treatment. Her husband, Howard Jones, was the gynecologic surgeon on the team. In 1954, the Joneses published a study showing that cortisone helped children with other hor-mone abnormalities, as well as congenital adrenal hyperplasia. (Cortisone gets converted to cortisol in the body.) A few years later, Howard Jones trumpeted Johns Hopkins's achievements in the treatment of intersex patients as a "therapeutic tour de force."

Perhaps the most influential member of this elite team was John Money, who provided advice to doctors and parents on how to treat children with ambiguous genitalia. He wasn't an endocrinologist or a surgeon or a psychiatrist; he wasn't even a medical doctor. He called himself a psychoendocrinologist and became the director of Johns Hopkins's new Office for Psychohormonal Research. He earned a PhD in social relations from Harvard in 1952 and wrote his doctoral thesis on the mental health of hermaphrodites (then the accepted medical term). Among his many findings was the supposition that hormones control sex drive but not sexual orientation. He also found that most of his subjects were surprisingly free of psychopathology, even though few of them had undergone any kind of medical therapy.

Money was an odd bird in a mainstream flock. He was prone to shockers. For instance, he proposed naming his field "fuckology." He showed pornography during his lectures at Johns Hopkins, claiming it would teach future doctors to be less judgmental when talking to patients about their sex lives. The students called his class "Sex with Money." He told them that he had developed a mathematical formula for the half-life of pornography, meaning that a viewer would become immune to the impact of an X-rated film after watching one for several hours.

He promoted some good ideas. For instance, at a time when doctors tried to use hormones to turn homosexuals into heterosexuals, Money insisted that therapy wasn't needed. In an article in the journal *Pediatrics*, he wrote, "There have been many illustrious homosexuals in the history of civilization. Some parents are reassured by this historical knowledge." He served as an expert witness in a widely publicized court case, defending the right of a homosexual eighth-grade teacher in Montgomery County, Maryland, to return to the classroom after he had been forced to take a desk job. (The teacher lost the case despite Money's testimony.) He also promoted bad ideas, among them that pedophilia was natural and should be accepted.

Money—who was married briefly in the 1950s—declared that monogamy in marriage no longer made sense because people live too long to remain sexually attracted to the same person. Unlike his media-shy colleagues, he courted the press and became a self-proclaimed sex guru. In 1973, he served on a panel about sexuality sponsored by *Playboy* alongside the porn star Linda Lovelace—not a very Johns Hopkins thing to do.

Money was not merely provocative; he harbored an unshakable devotion to his own theories. When he came to Hopkins, the thinking was that a person should be defined by their gonads: ovary equals girl; testes equals boy. That works most of the time, but not in all cases of intersexuality. Money held that the assigning of sex to a baby should not be based solely on gonads, solely on chromosomes (XX equals girl, XY equals boy), or solely on genital appearance. It should be an amalgam of all three—plus more. If the subject was a toddler, or older, their mannerisms, their dreams, and their sexual fantasies should be taken into consideration. He established seven criteria for assessing and treating children born with ambiguous genitalia:

1. Sex chromosomes (XX or XY)
2. Gonadal structure (are there testes or ovaries? are they plump or withered?)

3. Morphology of the external genitalia (penis too small? clitoris too big?)
4. Morphology of the internal genitalia (is there a vagina?)
5. Hormonal status
6. Sex of rearing
7. Gender role

The concept of gender role was the most novel of all. Money coined the term, coopting the word "gender" from grammar (where it refers to nouns in languages that use feminine, masculine, and sometimes neuter). Before this, the concept had been known simply as sex, a vague term that sometimes meant the act of intercourse, sometimes meant chromosomes, and sometimes meant feminine or masculine. "By gender role," Money explained, "we mean all those things a person says or does to disclose himself or herself as having the status of boy or man, girl or woman respectively."

The crux of Money's theories lay in the timing of treatment. He claimed that gender is formed by a three-step process, which involves hormones. During gestation, spurts of estrogen and testosterone wire the brain. After birth, you behave according to your brain wiring (feminine or masculine). That behavior elicits certain responses from those around you, who treat you like a boy or like a girl. Estrogen-drenched babies, for instance, acted girly and were therefore treated that way. These initial human interactions further shape one's sense of femininity or masculinity. Ultimately, the rush of hormones during puberty cements gender identity.

According to Money's theory, gender identity is malleable before the age of eighteen months—after the first hit of hormones to the brain, but long before the congealing act of puberty. It is about the time that people start to treat a child according to masculine/feminine norms, yet there is still the opportunity to treat a child differently. Before Money, hormones were thought to be the key determinant in gender and sexual orientation; Money stressed the importance of how a child is raised.

The Hopkins team used Money's new perspective, along with their own clinical experience, to develop treatment protocols. For instance, the team believed that boy babies born with the rare condition of micropenis should be made into girls. Howard Jones developed a surgical technique to mold a vagina from genital tissue. The testes were removed, and parents were instructed, as the Sullivans were, how to treat their one-time baby boy like a girl. Estrogen was prescribed at puberty to promote a feminine physique. The little boy could become a born-again girl as long as the treatment was started when gender was malleable—though the doctors were not dealing with someone born with ambiguous genitalia. They believed that if a penis was too little, the child would be happier raised as a girl, and that a girl with a too-large clitoris would be happier without one.

The doctors were not thinking about how prenatal hormones might influence gender identity over the long term; that was not considered until at least a few decades later. Nor did they conduct randomized controlled studies—that is, they did not divide children with ambiguous genitalia into two groups, those who had sex-switching surgery and those who didn't, and track them to see which group was happier over time. But they did observe the children and stay in touch with many of them after the surgical and hormonal treatment; they later reported that many of their patients lived satisfying lives.

Their guidelines—*the* guidelines—were devised to make a child feel as normal as possible. They believed they were supporting children's emotional stability. "There doesn't seem to be any doubt in the world but that living for many years with mixed-up-looking genitals is a serious handicap for anybody, whether they are living as a boy or living as a girl," Joan Hampson, a Hopkins psychologist, explained to colleagues at an American Urological Association conference in 1959. Surgery as soon as possible, she added, was "psychologically, extraordinarily important."

Except for one thing that Hampson didn't mention—that surgery was often not perfect, so afterward the genitals never really appeared as normal or functioned as well as the doctors and parents had hoped. Years later, in 2015, several United Nations agencies would condemn the practice of genital surgery on intersex infants. Malta was the first country to ban the procedure. In the summer of 2017, Human Rights Watch and InterACT, a group that advocates for intersex children, released a report lambasting genital surgery and pushing the U.S. Congress to ban it.

Back in the 1950s, as Elizabeth Reis, PhD, wrote in *Bodies in Doubt*, the Hopkins team provided reassurance and a concrete protocol where before there had been only confusion. Money's "bold articles were unusual in their conviction," wrote Reis, "and they must have come as a welcome relief to professionals confused by the many different solutions that physicians had put forth." Even so, there were dissenters. Milton Diamond, director of the Pacific Center for Sex and Society in Hawaii and author of the textbook *Sexual Decisions*, challenged Money's claims from the start. "I thought he was smart and I agreed with many of his attitudes and ideas," Diamond said recently, "but I certainly think he was wrong in the area of sex development." Diamond added that all too often physicians accepted Money's ideas without giving them careful thought. Money, he added, "had clout for all the wrong reasons." In 1997, a scathing scientific article by Diamond inspired an exposé published in *Rolling Stone* magazine and then in a best-selling book, *As Nature Made Him*, both authored by John Colapinto.

Diamond, and then Colapinto, detailed the life of a baby boy, a twin, who was made into a girl by the Hopkins team because of a botched circumcision. The baby was not intersex. According to Money's theory, the sex change should have worked because the doctors removed the tiny penis and instructed the parents to treat the child as a girl before he was eighteen months old. Medical reports called the procedure a success, but the child grew up depressed and confused—feeling like a boy, never fitting in, and not knowing why. He eventually learned about his medical history, reverted to living as a male, and ultimately committed suicide.

The article and book—published before the young man died—also accused Money of forcing the child to perform sexual acts with his twin brother. The outcry was polarizing. The Hopkins team lambasted the publications and defended Money until his death in 2008. But the public that had adored Money as a sexual libertarian and activist for homosexual rights now viewed him as a pervert. These days, most scholars see the good and the bad. Katrina Karkazis, PhD, a medical anthropologist at Stanford University's Center for Biomedical Ethics, wrote in *Fixing Sex* that, despite his flaws and despite the alleged abuse, Money provided for the first time "a nuanced analysis of the complexities of biological sex." He linked together, she added, the endocrinological, the psychological, and the surgical.

When Brian Sullivan became Bonnie Sullivan, it was a decision made according to the best practices of the time. The sex change was performed just before the eighteen-month deadline. The Columbia University doctors were confident that they were doing the right thing. According to the Hopkins protocol, Bonnie should have been another success story.

She wasn't. The day after Brian was erased, Bonnie stopped talking. No one knows why—not even Bonnie, who was asked years later. It's likely this small child was shell-shocked. No one called the toddler Brian anymore. Who was Bonnie? And what had happened to Brian's pants, favorite toys? Brian's world?

Bonnie was operated on again when she was eight. The doctor told her that she was going to have an operation to get rid of her bellyaches—though later, she didn't remember having any. They didn't tell her they were going to open up her abdomen in order to remove testicular tissue. She was admitted to Columbia University's Presbyterian Hospital on September 10, 1964, where she spent sixteen days in a ward with about eight other children with different problems. Because of her unusual condition and its teaching implications, a photographer took shots of Bonnie naked—a thin girl with a short bobbed haircut and fine features—and close-ups of her genitalia. There were lots of pre-operative exams. The fingers in her vagina and anus were mortifying. She felt like a freak. The other kids weren't getting examined the way she was. No one was photographing them.

Psychiatrists told her the surgery was considered a triumph. They reassured Ms. Sullivan that Bonnie was going to menstruate, have boyfriends, get married to a man, and make babies. But Bonnie didn't feel like the other girls. She was miserable.

When she was ten, her parents told her that her clitoris had been removed, but they didn't explain what a clitoris was. They said that it would have been a penis had she had been a boy, but since she had a vagina she didn't need one.

Bonnie started to recognize her same-sex desires in elementary school and figured she was condemned to a lonely, isolated life. She immersed herself in books, and developed an interest in computers when few people knew what they were. After starting and stopping her education, running away from home at one point and living with a group of hippies at another, she eventually earned a degree from the Massachusetts Institute of Technology.

Through it all though, she was plagued by her mysterious medical history. When she was nineteen, she went to the library and read up on sexuality and genital anatomy, including hermaphroditism. But that seemed too weird. She also read about DES, the hormone drug widely given to pregnant women to prevent miscarriages and later discovered to increase the risk of cancer and reproductive tract abnormalities in exposed babies. (This was the drug that Georgeanna Seegar Jones warned against.) Bonnie was convinced that she was a DES baby and that she was going to get cancer.

When she was about twenty and living in San Francisco—before attending MIT—she made an appointment with a gynecologist and pleaded with her to get her medical records. Columbia Presbyterian sent the doctor only three pages of what was certainly a large file.

"It seems your parents weren't sure if you were a boy or a girl," the doctor said, handing her the report.

Bonnie read: "Hermaphrodite."

And she read: "Sex of child doubtful, has both penis and vagina."

And she saw her birth name: Brian Arthur Sullivan.

"So I had these three pages, but I didn't talk about them with anyone. I was shocked and ashamed," she told me. The anger simmered. Later, she was plagued by suicidal thoughts.

In time, Bonnie got her full records, complete with the 1958 pathology report of her amputated clitoris. It said, "Part 1 is labeled clitoris, an elongated cylindrical penis-like structure which measures 3 cm in length." The three centimeters included not just the exposed part, but the internal part that was also removed. Some women retain slight sexual sensation if some of the clitoris remains after surgery, but not Bonnie. The report of the biopsy of her sex glands said, "ovarian tissue . . . testicular tissue . . . true hermaphrodite." And she read the nursing notes from her hospital stint when she was eight years old: "Quiet and uncommunicative. Helped clean up ward."

Bonnie found solace in feminist literature. In 1993, Anne Fausto-Sterling, PhD, a Brown University professor, wrote an article in *The Sciences* asking why children born with atypical features must be forced into one gender or the other. She somewhat facetiously put forth the notion of five sexes instead of two. She also criticized the practice of genital surgery on newborns, which could destroy a woman's sex life.

Bonnie was relieved to see a respected journal question the standard treatment for people like her and wrote a letter to the editor, which was published in the next issue, urging doctors to jettison the use of "hermaphrodite," which harks back to circus freaks, and replace it with "intersex." Bonnie did not invent the word "intersex"; it had been used interchangeably with "hermaphrodite" for years. Her mission was to eradicate the label "hermaphrodite."

Around that time, she started to open up to friends about her medical travails. Pretty soon people were saying that they knew someone like her, or had had an intersex lover, or had gone through the same thing themselves. In her letter to *The Sciences*, she asked people like herself to reach out to her so they could form a group, a place where they could share their experiences and relieve themselves of their desperate loneliness. She wanted to inform doctors managing intersex children that their interventions were largely misguided. In addition to her name, she provided an address for anyone who wanted to join the Intersex Society of North America. She rented a post office box specifically for the purpose, and used a pseudonym, Cheryl Chase, in order to protect her family. She chose the name by flipping through the phone book and had not intended it to stick, but it became her name for a while.

In truth, the intersex society didn't exist yet. It was just a sentence in her letter to the editor and a post office box. Within weeks, the box was jammed with handwritten letters divulging intimate details. The writers included their phone numbers. She spent her days responding to the letters and talking, for hours at a time, to people with similar issues to her own. They talked about loneliness and embarrassment. They had all sorts of questions about how their atypical hormones influenced their identities, their sexual inclinations. Some were as angry as she was about their surgical treatment and stints as naked medical models. Many experienced the same ongoing problems: no sexual sensation, endless complications. Many were on hormones for life.

Dr. Arlene Baratz was among the many people who reached out to Chase. "I really wanted to talk to someone else living with this," she said. Baratz knew about the history of intersexuality when few people talked openly about it. "It was heart-wrenching when these adult women talked about what happened in the 1950s and 1960s. They lived these lives in secrecy and isolation. Hearing these stories, I felt, this is not going to be my daughter's story."

In 1990, her six-year-old daughter, Katie, had been diagnosed with androgen insensitivity syndrome. During what was to be a routine repair of a hernia, doctors found a testicle. As a physician, Baratz understood exactly what the diagnosis meant and what to expect. She knew that her child had the XY chromosomes typical of a boy, but because her child did not respond to testosterone, she looked like a girl on the outside. However, she did not have ovaries or a uterus.

Children with androgen insensitivity do not menstruate, because they do not have a uterus, but they can grow breasts, because they often have enough estrogen. Katie, though, needed estrogen pills when she became a teenager, to carry her through puberty and for bone strength. "My emotions are similar to most: you grieve for your child's infertility," said Baratz. "And then I realized that the only thing she would not be able to do is have a biological child. I raised her with the understanding that might be okay."

Katie was interviewed by *Marie Claire* magazine and appeared on *Oprah* with her mother, and said that she wanted to become an activist for people like herself. She went to medical school and got a master's degree in bioethics. Today she is a psychiatrist. She is also married and a mother, thanks to an egg donor and a surrogate.

Unlike in the 1950s, doctors today are encouraged to speak openly with parents from the get-go. It was hardly a surprise that in 2013 Swiss and German researchers confirmed what most parents know: when doctors tell parents that there is no need to rush into a decision about surgery, they are more likely to delay or avoid operations compared to parents who are told it's an emergency situation. There are also, thanks in part to Cheryl Chase and others like her, several online and in-person support groups, so children and their parents no longer need to feel isolated. (Chase wasn't the first to launch a group, but most of them existed in secrecy.)

Debates about treatment continue. Brian/Bonnie/Cheryl Chase now goes by the name Bo Laurent. "Bo" harks back to the Bonnie days, and Laurent is for Laurent Clerc, a nineteenth-century student who fought for the rights of the deaf so that they would no longer be treated as mentally challenged, which was often the case. (Bo's maternal grandparents were both deaf; her mother's first lan-

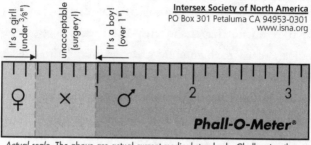

Actual scale. The above are actual current medical standards. Challenging these arbitrary standards, ISNA works to create a world free of shame, secrecy, and unwanted genital plastic surgery for children born with mixed sex anatomy.

An image created by Bo Laurent to express her anger about the surgery based on specific cutoffs. *Courtesy of Bo Laurent.*

guage was sign language.) "I wanted to do the same for the intersex people," she said.

Today, Bo lives with her partner in a cozy home in a quiet rural town in Sonoma County, in Northern California. She has a curvy physique and thick dark hair that hangs just below her shoulders. She remains in touch with a vast international community of patients and doctors. Her calm, soothing voice belies her inner rage about the way the medical community has treated people like herself. She equates genital surgery on intersex people with genital mutilation. The operations continue; Bo does not believe that much has changed since she was born. "It's true that people can now find others with intersex—but not thanks to their doctors—and doctors are less mendacious—but only, I think, because they know their patients and patients' parents will find intersex activists on the Internet." Intersex children and their families, she said, "are still subjected to additional harms, on top of their challenging situation of birth, and unnecessary suffering, imposed by a medical system that is resistant to change."

Bo Laurent can't erase the past—that's already been tried. I spent two days with her, as she shared intimate details of her life, her medical records, and an old leather album of baby pictures, the only one that was saved. Bo's aunt gave it to her after her mother died. In the photos (as in many baby photos), it's impossible to tell whether the child is a boy or a girl. In the bottom right corner of the leather cover is what must have once been a gold-embossed "Brian Sullivan," but the "Brian" was removed. In its place is a scratched, bald area, as when packing tape is peeled off a package and the top layer of cardboard goes with it.

Brian was born in 1956, at the height of the booming postwar economy, when people were leaving the cities for picket-fenced suburban homes and the American dream looked like a working husband, a stay-at-home mom, and two children: a feminine, pink-garbed girl and a sporty blue-suited boy. The Sullivans just wanted their child to fit in. Some scholars, looking back on those mid-century days, have suggested that the surgeries and hormone therapies solidified a binary gender system—that girls had to look one way and boys another. In many ways, the new scientific insights into hormones and how they influence development conflicted with long-held notions of maleness and femaleness. The data were showing a much more complex picture of humanity.

A decade after Bo went under the knife, in the 1960s, hordes of parents would seek medical therapy for another type of atypical child: those deemed too short. The parents wanted growth hormone, a cutting-edge treatment protocol supported by the science of the time. Unlike the parents of the 1950s, these parents were activists working alongside the doctors to champion hormone therapy. The goal was the same: to make kids who were different conform, and therefore make them happier.

8.

Growing Up

LIKE ANY SEVEN-YEAR-OLD, Jeffrey Balaban was annoyed by his annual physical. No kid likes to be poked and interrogated by a grownup. Jeff liked to think of himself as a healthy kid, but every now and then doctors would look at him strangely, curious about his height, or lack thereof. "He was obviously short," said his mother, Barbara Balaban. He was about 3 feet 5 inches then. "But we were of the notion that different people do things different ways and you don't make a fuss about differences."

On that day in 1960, after the pediatrician said that everything was fine, he asked if they wanted to talk to a growth expert. Ms. Balaban said no, not interested. And that was that.

Jeff was the shortest kid in his class, maybe the shortest in the entire grade. There were times when the top of his head barely reached a classmate's earlobes. His older brother and younger sister were short, but not that short. His parents were below average, too. Ms. Balaban hovered somewhere around 5 feet 2½ inches. Dr. Al

Balaban, her husband, reached somewhere between 5 feet 7 inches and 5 feet 8 inches. They thought the whole thing was silly. Drag the kid to a medical consultation for being short?

The next year's checkup was a rerun. Jeff's health was fine, but again the pediatrician asked his mother if she'd like to consult an expert in growth problems. This time, the doctor added that Jeff might never make it past 4 feet. Ms. Balaban was shocked. The doctor, she said, used the term dwarf, or midget, or some kind of 1960s'-vintage "little person" label. She knew Jeff was little; he was eight and about a half an inch shorter than his five-year-old sister. Still, she had never thought of his stature as a disability, or an illness.

Short is not a diagnosis; it's a description. But sometimes it is a sign that something is wrong. There are at least two hundred medical syndromes that stunt growth. A genetic defect can cause abnormal bone development, as in the case of achondroplasia, resulting in limbs too short and a head too large for a normal-sized torso. Or a lack of growth hormone may thwart the typical vertical climb of childhood. Doctors call these kids hypopituitary dwarfs, or hypopits for short. Sometimes kids are small because that is their normal—all the genes and hormones are up and running just fine. They just have small parents.

Every living being grows. We humans are unique in our pacing. Unlike other mammals, we hit the gas pedal right before birth and then gently press the brakes as soon as we hit the outside world. We still grow, but much more slowly. So we plump up before we get out and slow down afterward, relaxing into a drawn-out childhood. Think about it: our pets have reached their final size and are making babies when most of us are still on tippy-toes trying to sip from a water fountain. If, say, you get a German shepherd puppy about the same time you have a baby (as I did), you'll notice that by six months, your baby fits snugly in a Snugli. Your dog doesn't.

Anthropologists have theorized that we have a prolonged youth, compared to our mammalian cousins, in order to acquire greater amounts of knowledge to pass along to the next generation. Doctors see this through the prism of hormones. The difference between your little human and your little dog is, in part, the timing of the release of these chemical substances. When your dog is about six months and your child about twelve years, one brain gland (the hypothalamus) tells another one (the pituitary) to dial up growth hormone output. Vertical growth, scientists discovered, depends not just on growth hormone but on sex hormones too. Growth hormone also prompts the release of other helper hormones, which in turn stretch bones and muscles. A child who isn't growing may lack growth hormone. Or he could have the right amount of it but be lacking in the helper hormones. Or he could have all the right amounts of hormone, but signaling issues mean that his body isn't passing along the messages correctly.

The whole process of growing—limbs lengthening, muscles bolstering, internal organs expanding—depends on Swiss-clock timing and a master chef's attention to detail. It's like baking a cake: you can measure all the right quantities of eggs, sugar, butter, and flour, but if you don't know how to mix them together and you forget to turn on the oven, you'll end up with one sad dessert.

The early 1960s was an Alice in Wonderland time for scientists working on growth hormone.* They were supersizing laboratory

---

* Harvey Cushing saw it coming. As he said, decades earlier, "The Lewis Carroll of to-day would have Alice nibble from a pituitary mushroom in her left hand and a lutein [a portion of the ovary] in her right hand and presto! She is any height desired!"

animals and stunting others, and testing growth hormone, along with testosterone and thyroid hormone, on children. Jeff Balaban may have been just a small kid, a product of small parents, or he may have been deficient in growth hormone. There were no tests then to measure his levels. It was a matter of clinical judgment (did this kid look hormone-deprived?) combined with educated guesswork.

Years earlier, long before growth hormone was isolated, doctors had predicted that a hormone could "cure" shortness. Or, as Dr. Oscar Riddle, a past president of the Endocrine Society, told the Associated Press in 1937, because short people suffer from inferiority complexes, at some point in the future doctors will be able to give "runts" shots of growth hormone, thereby allowing them to reach their full potential intellectually and physically.

By the early 1960s, doctors were finally able to do something about shortness. Fueling the excitement was a flurry of articles in magazines and scientific journals repeating the age-old claim that short people, particularly boys, were condemned to an awful fate. A cover story in *Parade* magazine trumpeting the new hormone treatment referred to a "life of hellish dwarfism." John Money, the famous gender identity specialist, claimed that adults unintentionally coddle short kids, treating them as younger than their age, which encouraged immaturity and insecurity. Short kids, experts declared, were less likely to marry or land jobs compared to their lankier brethren. As Sheila and David Rothman write in *The Pursuit of Perfection: The Promise and Perils of Medical Enhancement*, "The combination of endocrinology identifying a state of hormone deficiency and psychiatry analyzing the degree of maladjustment suggested that shortness of stature was a disease, and by no means a trivial one."

The fear-mongering added to the urgency to treat. Perhaps the availability of new treatments biased the emotional studies, which proved that the newfound drug was needed. Were the researchers creating a therapy to treat an ailment, or did they concoct something and then search for a disease? Naturally, parents of short children, particularly boys, wanted to do whatever it took to secure for their kids a life of health, happiness, marriageability and hireability.

Hormone therapy was no longer the quackery of decades past, doctors insisted. This was modern medicine, with chemicals extracted in leading laboratories, the ingredients measured, and the children monitored.

The Balabans and their three children lived in Great Neck, New York, a Long Island suburb about forty minutes' drive from Manhattan. Barbara Balaban worked in various editorial and secretarial jobs and volunteered at the public school and in the community. Al Balaban was a psychiatrist, and while he kept up with all things medical, he hadn't heard much—or at least thought much—about growth treatments. Sure, the newspapers sporadically published stories about growth hormone breakthroughs. Chemists mined growth hormone from a cow's brain; that made headlines in 1944. Scientists deciphered the structure of growth hormone; that made headlines a decade later. In 1958, newspapers wrote about a cure for dwarfism with growth hormone, and trumpeted the possibility of tailor-made hormone therapy in 1959. But with this breakthrough and that one, and so many other things going on in the world (race riots, space missions, Vietnam, and so on), who remembered every medical discovery? Unless, that is, you thought it related to you. For the Balabans, height became an issue at Jeff's eight-year-old checkup.

When the doctor asked again if Ms. Balaban wanted an expert opinion, she worried that she wasn't worrying enough. Maybe she thought about those newspaper articles. Even though she had considered Jeff just fine the way he was, she now considered the possibility that he wasn't. Surely, a consultation couldn't hurt.

The Balabans had a hunch that Jeff would fight against testing and treatment and labeling. But they believed he would be grateful in the long run. "It was a bad time for him. He was caught in these cross currents," said Dr. Balaban. "He had grown up being cute. He was adorable as a young child—garrulous, sociable, full of fun, and people took to him. He was mischievous as hell. He was being treated by his teachers as cute and allowed to do things he shouldn't be allowed to do. But he was tormented by the kids on the playground. He was liked but abused."

One afternoon in 1961, Ms. Balaban pulled Jeff out of school to see Dr. Edna Sobel, a pediatrician at the Albert Einstein Hospital in the Bronx, an hour's drive from their home. Dr. Sobel specialized in children with hormone problems and was considered one of the leaders in the field. She had trained at Harvard and was involved in several landmark studies of dwarfism. Though she was known by some of her colleagues and patients as a compassionate physician, Ms. Balaban remembered Dr. Sobel as gruff and curt. She remembered that Dr. Sobel was often in a wheelchair. Sometimes she would be standing up, hunched over. She appeared small, smaller than Ms. Balaban. Though Dr. Sobel never mentioned it to her patients, she had had polio as a child and the boney deformities caused by her illness triggered growth problems. She wore a lift in one of her shoes and was in chronic pain.

"The workup was expensive and it went on forever," said Ms. Balaban. There were blood tests, and everything was measured. Jeff wasn't happy. He hated missing school. He hated feeling like a freak. He hated getting naked in front of a woman, even if she was a doctor.

Dr. Sobel suspected that Jeff had hypopituitarism, meaning that his pituitary wasn't producing enough growth hormone. Dr. Balaban was shocked. "I had heard about it in medical school, Tom Thumb sort of stuff," he said, referring to the 3-foot-4-inch dwarf who appeared in circuses in the late 1800s. He had never put his son in that category. Jeff was a normal kid, not a circus freak.

For doctors, the chief concern isn't stunted growth, but all the other things growth hormone does. It helps to balance sugar, to metabolize proteins and fats, to maintain heart and kidney health, and to stimulate the immune system, to name just a few of its many tasks. So, growth hormone isn't just about growing, and growing isn't just about growth hormone. A better name for this chemical may be growing-up hormone.

Dr. Sobel suggested thyroid hormone first, since boosting Jeff's metabolism might help him grow. A few years earlier, Jeff might have been given testosterone shots. But testosterone, doctors realized, made boys grow sooner, but not taller than they would have been without it. It was like traveling by high-speed train: faster, but you end up at the same place. In fact, it was Dr. Sobel who authored one of the key studies showing that testosterone didn't increase growth.

The Balabans agreed to try thyroid treatment. A few months and many shots later, Jeff reluctantly left school early again and endured the drive from Long Island to the Bronx. He fidgeted impatiently with other short kids in the crowded hospital waiting room; the norm was a two-hour wait for a fifteen-minute appointment. When Jeff's name was finally called, Dr. Sobel told him what he had already figured out on his own: the thyroid shots weren't working. No growth spurt at all. In fact, he hadn't grown even a little.

If Jeff had gone to the doctor just two years later, he would have had several blood tests to measure the ups and downs of his growth hormone. But those tests hadn't been invented yet. Instead, he was admitted to the hospital for a month. Doctors measured his food intake and his output. They X-rayed his parents' heads to see if they had any pituitary issues that might have been passed along to Jeff. There were no sophisticated brain imaging machines in 1961, so doctors gleaned clues about the pituitary by X-raying the boney dish where the gland rests. Crunched or splayed bones signaled trouble, because the distortion suggested a tumor—not proof, but a clue. Jeff was given a special X-ray procedure called a pneumoencephalogram, a mouthful of a term that describes a seemingly barbaric procedure devised in 1918 and used until the 1970s, when gentler methods of seeing the skull finally came along. Doctors drained the fluid around Jeff's spinal cord, pumped air into his head, and then X-rayed it, which gave a clearer view. Jeff went home writhing from an excruciating headache, an after-effect that was par for the course.

The tests showed he did not have a tumor. Dr. Sobel presented one more option: human growth hormone. This therapy was brand-new, and initial reports were promising. Kids grew. There were no comparison studies, but for children who seemed to be deficient in growth hormone, it made perfect sense. Give them what they seemed to lack. Normalize them.

For the Balabans, switching from thyroid, which didn't work, to growth hormone, which might work, was an easy decision. The difficult step had been starting shots in the first place. That's when Jeff had turned from person to patient. That's when his life gained a before and after: before shots and after shots. Before, when short was a description, to after, when short was a diagnosis. Once the Balabans had stepped over the murky line separating healthy and unhealthy, they were more willing to try other medicines.

But there was a difference, a huge one, that Ms. Balaban didn't yet know about. Thyroid hormone was plentiful. Growth hormone wasn't. Growth hormone therapy for children was like a gold rush among doctors, each racing to get their share of the precious extract. When Ms. Balaban agreed to try growth hormone, Dr. Sobel laughed. That rattled Ms. Balaban. Was she mocking her? Was it all a sick joke? Albert Einstein Hospital had a small stash of growth hormone, Dr. Sobel explained, but it was promised to another child. Dr. Sobel had laughed because she was surprised that Ms. Balaban assumed it was an easy fix, as if the growth hormone were stocked in a medicine cabinet. It was anything but.

Growth hormone for treating children was obtained from the brains of dead people. Hence the name "human growth hormone," instead of plain "growth hormone." One pituitary supplied just enough to treat one child for one day. Doctors assumed short children needed a shot every day for at least a year. There had been no studies testing dosage, but for whatever reason, the contents of one pituitary seemed like the right amount. That meant 365 pituitaries, or 365 dead bodies, to make one child grow. You don't have to be a mathematician to realize that treating all the severely short kids in the country, perhaps many thousands of them, equaled morgue-fuls of cadavers.

Dr. Sobel then made one of the oddest proposals a doctor would ever make to a patient. She told Ms. Balaban that if she wanted this scarce drug for her son, she'd have to collect her own batch of pituitary glands. "She just looked at me and said, 'Do you know a pathologist or someone at your hospital? You'll need 100 grams and when you get them, we can process it.'"

If Dr. Sobel had asked Ms. Balaban to raise money, that would have been easy. If she had asked Ms. Balaban to stage a march through Washington, that would have been feasible as well. But she was talking about collecting a body part, one that was tucked deep inside the brain, no less—something you would think should be off-limits to a non-medical person.

Standing in the doctor's office with her son, Barbara Balaban knew she was about to embark on a long, strange journey. She just didn't know how strange. Within weeks, she and her husband would be traipsing around morgues across the country, and penetrating the inner sanctums of elite medical meetings. She would transform from a worried mother into one of the nation's leading pituitary bundlers. It took a little luck, some good connections, and a lot of moxie. Or as she put it, "We all did it out of sheer desperation."

In 1866, Pierre Marie, a French neurologist, figured out that a giant is a giant because he has an enlarged pituitary. It would be another half century before doctors pinpointed the precise chemical among the gland's many substances that triggered growth. The competition among investigators was like rival crews diving for sunken treasure. Everyone is swimming in the same area but only the first to grab the bounty reaps the prestige and the rewards.

The growth hormone honor would go to two scientists at the University of California, Berkeley. Dr. Herbert Evans, a former student of Harvey Cushing, and Dr. Choh Hao Li, a biochemist, announced their victory in a 1944 article in *Science*. Evans and Li began with chunks of pituitary, to confirm that it did indeed hold the necessary ingredients. They fed gland bits to rats and watched the rodents balloon. They removed the rodents' pituitaries and saw them shrink. Then they injected the gland into the rats again. The rats regrew.

It wasn't long after that study that the two scientists isolated the elusive growth hormone from the chunks of pituitary. A few skeptics doubted that they had gleaned pure growth hormone, and claimed what they'd found was a blend of thyroid and ovary and testicle. In other words, they suspected there was no such thing as a growth hormone per se, but rather a pituitary hormone that had many effects on the body. Evans and Li defended their position, concluding in their *Science* article that "a dose of 5.0 mg of the product did not show lactogenic, thyrotropic, adrenocorticotropic, follicle-stimulating or interstitial cell stimulating activities." This proved it was growth hormone, as it had no other "biologically active pituitary contaminants."

The Evans-Li study that grabbed the media spotlight involved two puppies. The scientists got a cow head from a slaughterhouse, retrieved growth hormone from its pituitary, pulverized it into a fine powder, and injected it into a dachshund puppy. The pup towered over its littermate. The canines no longer looked like siblings; the one given growth hormone not only grew bigger, but its neck thickened and its jaw spread. Under a photo of the dachshunds splashed across a page in *Life* magazine, the report said the larger pup looked more like a bull mastiff. Indeed, the experiment showed that growth hormone—as doctors with acromegalic patients had suspected—not only increases height but also triggers facial changes. These were the kinds of changes that Cushing had noted years before, which prompted him to write that letter to *Time* magazine lambasting the "Uglies" contest.

At first, doctors thought that an ample supply of growth hormone could be retrieved from animals. The thinking was that growth hormone was growth hormone, regardless of where it came from. If cow hormone worked in rats and dogs, it would work in humans. Look at insulin: it came from pigs, and it controlled humans' blood sugar swings.

Unfortunately, growth hormone didn't work in the way that insulin did. Pig growth hormone made mice grow, but it didn't add inches to people. Doctors who treated patients with bovine growth hormone found that it didn't do anything.

In 1958, Tufts University's Dr. Maurice Raben announced that he had made a dwarf grow using growth hormone from a human cadaver. His brief, dry letter was tucked away on page 901 of the August issue of the *Journal of Clinical Endocrinology and Metabolism*. He reported that he gave his patient one milligram of human growth hormone twice a week for two months, then 2 milligrams three times a week for seven months, boosting the patient's height 2.67 inches.

Raben had been competing with the Evans lab in California, and other labs, to be the first to report success with human growth hormone. (His lab had already lost when it came to isolating animal growth hormone.) Raben submitted his findings in a letter rather than as an article. With a little promotion, letters were just as capable of capturing the media's attention. The Tufts hormone conquest made headlines. "Hormone Makes Dwarf Grow: May Also Offer Clues in Cancer, Obesity and Aging," ran one in the *New York Herald Tribune*.

The news that humans can only use human growth hormone was exciting, but a few doctors saw the potential for abuse. It "won't produce basketball players par excellence," quipped Philip Henneman, a Harvard physician. Most people didn't appreciate that the discovery meant that supplies would be limited.

Jeff Balaban was back in Dr. Sobel's office at the beginning of the school year in 1961, two years after the dwarf-cure hullabaloo. Ms. Balaban was told that her son would need three shots a week. Ideally, patients were to be given a shot a day—a dose calculated on feelings rather than facts. But supplies were too limited for that, so three times a week would have to do.

That meant, for Jeff, 156 pituitaries, or 156 cadavers, for a year's supply. That's a lot of corpses. At the time, the Balabans had no idea that Jeff would be treated for the next ten years; it would take a graveyard of bodies to make him grow. Dr. Sobel said that if Ms. Balaban could harvest 100 pituitaries, they would start treatment. "We called our best friend, who is a surgeon, and another friend, a pathologist, but they said they were already committed to another program," said Ms. Balaban, referring to another growth hormone collection organization.

Looking back on it all decades later from their south Florida retirement community, the Balabans have a hunch that Dr. Sobel gave them the option to collect their own hormone not so much because she really thought they would obtain 100 pituitaries but because she didn't want to tell them it was hopeless. She "looked at us ruefully and said, 'Sorry, fellas, it's not available,'" said Dr. Balaban, but after a moment of silence she added: if you know someone who knows someone, maybe we can help you.

As they told it, that one flicker of hope from Dr. Sobel—even if it wasn't truly hopeful—mobilized Ms. Balaban and depressed Dr. Balaban. "The two of us sat around and cried for three days," said Ms. Balaban. "And then I got angry and at the bottom of it all is that we are responsible for this kid and we have to do anything and everything we can so if he ends up at 4 feet tall, we can say we did everything we could."

Maybe it would have been better if a cure hadn't been dangled in front of her. If they had never agreed to the consultation, they would never have started on a path of an experimental, out-of-reach therapy. But the die was cast.

Barbara Balaban felt she couldn't deny her son something that could make him happy: being taller. Something other short kids had access to. So she did what she did best: she launched a grass-roots campaign. What worked for the Parent Teacher Association and the draft board would work for pituitaries, she believed. "We were all encouraged to volunteer," she said, "and I was lucky to have a husband who didn't mind my spending money on meetings and hosting people at our home."

Dr. Balaban figured he would contact a medical school class-mate who had gone into pathology. His wife said that wasn't enough. They had to write letters to everyone they knew, asking if anyone knew anyone who knew a pathologist willing to donate a pituitary. Dr. Balaban looked at her and said, "What are you trying to do, start a national organization?"

That is exactly what she did. Ms. Balaban went on to found the Human Growth Foundation. Its purpose was to inform families about getting treatment and coping with a diagnosis of dwarfism. And the Balabans would, in time, become founding members of the National Pituitary Agency. But on that day in 1961, they were just thinking about one pituitary at a time.

Ms. Balaban sat at her kitchen table and typed letters to every-one she knew. "When I say everyone, I mean everyone Al had gone to medical school with and everyone on every committee I'd ever been on, every parent of every child in all of my three kids' classes." It wasn't easy in the days before email and all the other ways to cyber-blast messages. The letter told of the Bala-bans' desperation. They needed pituitary glands to save their son from what they considered to be a life of miserable smallness. They urged friends to contact hospitals and to spread the word in their schools and churches and synagogues. She told them that the pituitaries should be put in a tube with acetone, nail polish remover, which preserved them. She mailed the first batch of letters in November 1961.

Someone called about one gland. Then someone called about another. One caller had three. "We were ecstatic. We went around picking them up. I got a call from a friend one day and she said, 'I've got a gland for you.' And I said, 'Where in the world did you get it?' She said she was at a wedding and the father of the bride slipped her this package addressed to me." Ms. Balaban said each package that arrived at her door looked like a jar of peas. To paraphrase Dr. Salvatore Raiti, a leading endocrinologist, a thousand pituitaries could fit into a half-gallon milk container.

For the most part, pathologists would bottle the brain glands in acetone and hand them off to the Balabans or a friend. Typically, Dr. Balaban picked them up from the pathologist's office, but sometimes he went to the morgue. Then the Balabans emptied the vials into larger glass jars filled with fresh acetone and stored them in a closet in the laundry room.

In those days, anyone could plunk a pituitary into a mason jar of nail polish remover and do whatever they liked with it. You could hand it off to a needy parent. You could pop it in the mail. Some pathologists froze their brain glands, which yielded more hormone than non-frozen ones, but if they thawed accidentally (let's say you hit a traffic jam), you lost the whole thing. Forty years into the future, these same brain bits would be classified as a biohazard, necessitating permits and precautions when they are moved—not to mention that you now need the permission of family members to give away a body part from their loved one.

"We didn't think about whether it was legal," said Ms. Balaban. "It was the days before HIPAA [the patient privacy act]. The glands

Jars of pituitaries. *Ralph Morse/The Life Picture Collection/Getty Images.*

were only available on autopsy. Families weren't giving permission. We just did it and never thought about any of those aspects.

"What happened was everybody referred us to somebody else and one day, I got a call from a guy who said, 'I have three glands,' and I said, 'I'll come right over.' He said, 'You're planning on coming to Texas?' So he said he would mail them to us. He mailed us this cylindrical container wrapped in cardboard and inside batting to protect the vial and inside the vial was acetone and three glands and we looked at each other and said, 'That's how it's done.' So we went out and got stuff.

"And that's when we started sending mailing packages with our letters. We got a vial with a screw top and cotton, mailing tubes, self-addressed labels, and lots of postage stamps and wrapping paper and sent them to anybody who could get us glands. It didn't cost them anything. Whenever people sent us stuff, we'd send them more packaging for the next one."

Ms. Balaban kept a record of everyone who donated a pituitary, everyone who referred her to someone else, everyone who knocked on doors to ask for help, on three-by-five cards. The cards were alphabetized and color-coded: green for active supplier, red for active referrer. Everyone got a thank-you note.

The Balabans spent that Christmas with friends in New Jersey. When they came home, there was a little package in the mailbox: the last few glands they needed to reach 100. "It took other people six months to get to 100," said Ms. Balaban. "We did it in a month."

She returned to the hospital in the Bronx with her first batch of 100 pituitaries, assuming that Dr. Sobel would be ecstatic and Jeff would start treatment shortly thereafter. Instead, Dr. Sobel was astonishingly nonchalant. "My suspicion as I got to know her a little bit and saw her full-page ads against the Vietnam War and Agent Orange was that she was thinking that we, the privileged, could do that [have access to treatment]. And the kids she was treating in this city hospital didn't have the resources or the experience in life."

Even more stunning for Ms. Balaban was when Dr. Sobel said the Balabans would have to wait at least three months while the glands were turned into a hormone treatment.

There were three labs in the country extracting growth hormone: at the University of California, Berkeley; Tufts University;

and Emory University. Purifying the hormone was in some sense like honing a gemstone from a chunk of rock. It took persistence, care, and dexterity. Each lab developed its own technique to obtain the cleanest product. The Balabans' pituitaries went to the lab of Dr. Alfred Wilhelmi at Emory, on the understanding that Ms. Balaban got half for her son and Dr. Wilhelmi kept half for his own research. She had no choice; she had to get someone to extract the hormone and every extractor insisted on keeping a portion of the stockpile.

Jeff hated the whole thing from the start. His father gave him the shots, and they stung. "I remember the look of anguish on his face," said Dr. Balaban. But they believed they were doing the right thing for him. Dr. Balaban told Jeff that he had no say in the matter until he was older and could appreciate the implications of the treatment, or rather the implications of *not* getting it.

Every month required another day off school for Jeff to check in with the doctor. He'd lie there stark naked as she measured every body part. "They measured his penis. It was awful," said Ms. Balaban.

"One day, about a year or so into treatment," she said, "this guy comes looking for us. Someone the government sent. And he tells us we are the third largest collectors of pituitaries in the country, exceeded only by the National Institutes of Health and the Veterans Administration. They wanted us to release our resources. Everywhere he went, he said, he would ask a pathologist to join the government program and he'd be told they were already committed to Balaban. He didn't know what Balaban was. So we gave him the same mandate we gave everyone else." The Balabans would share their stash as long as there was always enough for Jeff.

Dr. Robert Blizzard, a pediatric endocrinologist who had

done many of the original growth hormone experiments, had been collecting pituitaries for patients at Johns Hopkins Hospital, offering pathologists two dollars a pop. The Balabans didn't pay pathologists.*

The competition among clinicians to reap their share of the pituitary bounty was starting to get ugly. Some doctors upped the price to pathologists, hoping to secure more pituitaries. Dr. Blizzard worried about a black market, one where only the pushiest parents or the ones with the most money would get treatment. In 1963, he organized a meeting of the largest extractors, along with scientists and a few parents of short children. The Balabans were there.

Dr. Blizzard proposed they pool and share resources. The largest pituitary bundlers worried that a central collection facility would diminish their supplies, so Dr. Blizzard suggested that no one get fewer pituitaries than they had been getting before the collaboration. They called themselves the National Pituitary Agency and launched in 1963, sponsored by the National Institutes of Health. The agency was first run out of Hopkins, with Dr. Blizzard spearheading, and was later directed by Dr. Salvatore Raiti at the University of Maryland.

Because the National Institutes of Health funds experiments, not treatment, anyone who got pituitaries through the National Pituitary Agency had to be part of a scientific study. The treatment was crucial, it was thought, so nobody in the study would be given a placebo; being part of a study meant merely that you were

---

* Gil Solitaire, a retired neuropathologist who was at Yale during the pituitary drive, remembers shipping pituitaries and anything else of interest to Hopkins, although he doesn't remember getting paid. "I just knew that if we had a pituitary, send it to Hopkins. And if we had a brain with anything interesting, Hopkins wanted half of that, too. So I used to like to say, if you want to get into Hopkins, all you need is half a brain."

monitored thoroughly and your medical information was kept in a registry, albeit anonymously.

National Pituitary Agency doctors did all they could to publicize their cause, and also to prevent non-NPA folks from encroaching on their territory. They pressured foreign-based drug companies to collect pituitaries from outside the U.S. They published a newsletter urging anyone who had access to a dead brain to contribute to the cause. They encouraged journalists to write articles about dwarfism and the need for pituitaries and tried (unsuccessfully) to incorporate storylines about hypopituitarism in TV shows, such as *Dr. Kildare* and *Ben Casey*. They solicited help from anyone with connections. Fred Mahler, a pilot for TWA, had two children with hypopituitarism. (His other two children did not have it.) Mahler agreed to pick up and deliver pituitaries for free, placing the batches near him in the cockpit. His organization, Pilots for Pituitaries, grew to include six hundred doctors and fifty pilots. In 1968, at a meeting of the College of American Pathologists, where Mahler was honored for his work, he said he wanted to help the pituitary agency because "otherwise, it would have been jungle warfare, with each parent trying to get what his child needs."

The agency also issued guidelines. For instance, they advised that hormone shots stop when boys reach 5 feet 6 inches and girls reach 5 feet 3 inches. The concern was not about overdosing or staying on hormones for too long, but rather about sharing the scarce drug, so that every child could have an opportunity to reach an adequate height.

In the meantime, biotech companies were trying to figure out a way to make growth hormone from scratch, avoiding the need for cadavers and boosting supply. But many clinicians worried about synthetics and considered the brain-derived hormone the natural, safer choice. Potency varied batch to batch. In the early days, doctors tested each sample by giving a tiny amount to a rat that had had its pituitary removed and then waiting a few weeks to see if it worked. It was crude, but it was the best method available.

Despite the push for treatment, despite the excitement, and despite sensationalized newspaper headlines such as "We Can End Dwarfism," there was never any guarantee that growth hormone—even one milligram a day for at least ten years—would work. There were no experiments comparing those taking the drug to those not taking it. For some children, it seemed to work, stretching them from under four feet to perhaps an even five feet or a few inches beyond. Others saw no effect at all. Either way, it was impossible to know how much the child would have grown without it.

Jeff Balaban got shots three times a week from the age of eight to the age of seventeen. He reached 5 feet 3 inches—which his parents believe was due to the growth hormone. Perhaps he would have gained a few inches anyhow. Jeff hated the whole rigmarole not least because it was a constant reminder that he was different—that he didn't fit in. On July 8, 1971, Jeff announced that he was finished. His parents agreed that he was old enough to understand the consequences. Though their son had quit the program, the Balabans remained involved with support groups for parents of children considered dwarfs.

For a while, the collection and distribution of growth hormone seemed to be running even better than expected. By 1977, extraction was centralized at one lab: Dr. Albert Parlow's, at the University of California, Los Angeles. Dr. Parlow was able to harvest seven times more hormone from each pituitary than other extractors managed. More importantly, Dr. Parlow—who had been a young scientist in the early days of hormone purification—believed that his experience and his obsessive meticulousness yielded the purest product.

Pituitaries from all over the nation funneled to Los Angeles and fanned out back across the country to children in need. It was a system built on the complex coordination of volunteer parents, pediatricians, biochemists, and endocrinologists. It seemed, for a fleeting moment, to represent the best of American medicine. Until data proved otherwise.

9.

Measuring the Immeasurable

I N THE 1970S, a peculiar ailment struck one in 4,000 children. Their heads were too large, their necks too fat. Their skin was scaly and dry. Their tongues were thick and flabby and drooped over their chins like a wilting flower. Mothers worried because their infants, though pudgy, barely ate and were floppy as rag dolls. As the children got older, further disturbing symptoms appeared. They fumbled to find the right words. They could barely manage the journey from spoon to mouth. Just looking someone in the eyes was a struggle. There was a name for these children: doctors called them cretins. The word soon became slang for "stupid," or "idiot."

Strangely enough, a cure had been known for nearly a hundred years. Doctors knew what triggered the ailment: a deficiency in thyroid hormone. And they knew how to raise the hormone level: thyroid tablets, which were easy to obtain and cheap. The drugs revved up metabolism. Newborns could be fed pills that dissolved in water, formula, or breast milk. Still, children suffered—because the disease could only be staunched if it were picked up at birth. And because the babies looked healthy when they were born, most of them defied detection. By the time doctors saw the telltale signs—often not until the child was six months old—it was too late. Thyroid pills couldn't reverse the brain damage, once done.

In the 1980s, when I was in my third year of medical school—when we finally went into a hospital to learn from real patients, not just images and descriptions in textbooks—a professor arrived one day accompanied by a woman he described as a cretin. She had been invited, or perhaps cajoled, to spend an hour or so talking with us in a cramped conference hall. She was in her twenties, about my age, and was stocky and round-faced, with short-cropped brown hair. She was smiley and shy. I don't remember the conversation, just that it was awkward. She seemed to feel special, as if she had been invited as an expert speaker, which in a way she was. She was there to teach us: to show us that she was the way she was because someone had made a mistake, decades earlier, missing the diagnosis when she was born.

Nowadays we rarely hear about cretinism. Doctors no longer parade such people on the pediatric wards. Millennials may likely not even know the word. The affliction has been wiped out—at least in the developed world. That success is due to a technology invented by a little-known but very important scientist—a woman from the Bronx who devised a way to measure the immeasurable.

Rosalyn Yalow was an unlikely success story, a Jewish woman who grew up at a time when Jews had restricted access to major institutions and women were often prohibited altogether. And yet, almost everyone in the world has had an affliction whose treatment has been informed by her work.

The second of two children, Yalow was born on July 19, 1921, to poor immigrants from Russia. Her parents did not graduate from high school but were voracious readers, trying to keep up by reading their children's schoolbooks. Yalow was raised to make do with little—a lesson that would come in handy years later, when she was given a closet for a laboratory and little funding, but made her extraordinary breakthrough anyhow. When she was eight years old and the already tight family finances constricted even further, her mother took on work at home sewing collars into men's shirts. Yalow's role was to stretch the fabric taut so her mother could stitch. As her biographer noted, Yalow knew from an early age how to "bear down, push out trouble, and focus on the work."

She attended the local public high school and then Hunter, the local public college, graduating cum laude with a degree in physics. She wanted to be a scientist, but her teachers suggested she become a scientist's secretary. Disheartened but not ready to abandon her dreams, Yalow took a secretarial job for a Columbia University professor of biochemistry, hoping that as an employee she could take classes. Yalow was thinking science. The professor suggested stenography.

Yalow almost succeeded in being admitted to Purdue University as a graduate student. "She is from New York. She is Jewish. She is a woman," the admissions officer wrote to one of her Hunter professors. "If you can guarantee her a job afterward, we'll give her an assistantship." No guarantee was forthcoming, so Purdue rejected her, not wanting to waste a place on a student without job prospects. Yalow eventually made her way into a graduate program only because so many men were away fighting in World War II. "They had to have a war so I could get a PhD and a job in physics," she would say years later, showing a glimmer of dark humor. She nabbed the one opening for a woman at the University of Illinois's

College of Engineering. As soon as she opened her admissions letter, she threw her stenography books in the trash and headed west. Within her first days there, she met Aaron Yalow, a fellow student. They married the following year.

One day when her coursework was nearly completed, the department chairman called her into his office. She had gotten nearly all As. He pointed to her sole A-minus and told her, "This confirms that women don't do well in laboratory work."

Yet Yalow was on her way to developing one of the most crucial innovations in twentieth-century medicine. She completed her doctorate in 1945, a year before her husband, and headed back to New York. She wanted to work in a university-affiliated nuclear physics lab, but didn't get any offers. She accepted a job as temporary assistant professor of physics at Hunter College, but considered the post subpar because it was a women's college that didn't take physics seriously. (Hunter became co-educational in 1964.) Yalow nurtured her students, particularly encouraging the few students interested in science. She wasn't training the next generation of secretaries. "She pushed me into the wider world. She always said . . . you shouldn't give up anything," Mildred Dresselhaus, one of her former students, said. (Dresselhaus became the first female physics professor at the Massachusetts Institute of Technology, and had her moment of wider fame when she starred in a 2017 General Electric commercial touting women in science that showed paparazzi and teenage girls swooning over the eighty-six-year-old researcher as if she were a pop star.)

Aaron Yalow, who got a job teaching physics at Cooper Union in Manhattan, was the warmer half of the marriage, nurturing his wife's career and cultivating a community of neighbors and friends from the local synagogue. Rosalyn wasn't one for religion, but agreed to keep a kosher home for Aaron's sake. She cooked dinner every night and filled the freezer with individually wrapped home-made kosher meals for the times she was off lecturing or attending scientific meetings, which was often.

While teaching at Hunter, Yalow reached out to Columbia University physicists, throwing her name in the hat for any laboratory jobs that might come along. Her networking paid off. When the Bronx Veterans Administration started a nuclear medicine department, they called Columbia, who suggested Yalow. In 1950, she showed up at the VA in her new job, excited about the opportunity but annoyed that the lab was nonexistent. She turned what had been a janitor's closet into a laboratory.

Hardly any female scientists were hired by the VA, and the few who were had to quit if they got pregnant. Yalow refused. "When it came time for me to have children," she told Dr. Eugene Straus, her biographer, "I was too important to be fired. The rule at the VA was that you had to resign in the fifth month of pregnancy. Resign. Not get a leave. I always teased, saying that I was the only person to have an 8-pound-2-ounce five-month-old baby."

Her life was work and family, and they often blurred together. Her colleagues were invited to eat with the Yalows and go with them on vacations. Her children spent weekends at the lab. She'd go there every day and often after dinner, even on Saturdays. VA rules prohibited children in the lab, so as they drove by the gates of the main entrance Yalow would yell, "Duck," and the kids would crouch in the back seat until they got through security. Then they spent the day with the rodents or did their homework while Mom conducted experiments.

At the VA, Yalow connected with Solomon Berson, an internist eager to do research. The meeting was supposed to be Yalow interviewing Berson, a clinician with little research experience. But rather than she asking him questions, he challenged her to one math riddle after another. His intellect and bravura astonished her, and she hired him on the spot. He was thirty-two; she was twenty-nine. That first meeting sparked a lifelong friendship and partnership, an intellectual bond—a match made, if not in heaven, then in some scientific equivalent.

Yalow had no hobbies and little tolerance for those who were not intellectually up to snuff, which limited her friends to a small group of scientists. She had one soft spot: her lab rodents. She caressed them as she fed them each morning, and refused to kill them after the experiment was finished—the standard exit strategy. Instead she kept them, so her home became the refuge for a constant stream of guinea pigs and rabbits.

The kernel of the idea that led to her landmark contribution to medical science began with basic endocrinology studies. The thinking at the time was that hormones were so scarce they were impossible to measure. Also, along the same lines, doctors assumed when they treated their patients with hormones, they didn't have to worry about an immune response. When something foreign (a transplanted organ, say) enters the body, there is usually some kind of attack response from immune cells. Insulin, for instance, was harvested from animals in those days, and though most animal products would provoke an immune response, the thinking was that hormone therapy—because the hormones were administered in such tiny amounts—would not.

Yalow and Berson proved that notion wrong, by showing that many patients did mount an immune response to the therapy. Despite the scrupulous methodology described in the scientific paper that detailed the study, it was rejected by two leading publications: *Science* and the *Journal of Clinical Investigation*. The study was valid, as confirmed by peer reviewers who checked their methodology, yet the journals' editors refused to believe it.

Yalow wrote irate letters to the editors. insisting that the team's data had shifted the paradigm. Eventually, the *Journal of Clinical Investigation* agreed to publish the article, on condition that the duo agreed to delete the word "antibody." An antibody is a particular immune substance; while Yalow and Berman had proved that the body did indeed create antibodies to insulin treatment, the editors just couldn't accept it. They insisted that the nonspecific word "globulin" be used instead—akin to a meteorologist, hesitant to call something a tornado, just calling it a major wind. Yalow and Berson reluctantly agreed to the editor's semantics. The article was published in 1956, and quickly thereafter other labs confirmed their findings.

The insulin antibody work convinced the workaholic duo that they were onto something even more revolutionary. The driving question was how to measure these tiny amounts of insulin. Though common knowledge said hormones weren't measurable, surely there was a way. They combined their individual expertise in physics and endocrinology to devise a solution. The tool they invented was based on the fundamental principles according to which chemicals in the body bind to one another. Biology teachers like to say that when one chemical attaches to another, they join like a lock and key. One key fits into one door. One hormone latches onto one kind of immune cell. The couples are destined.

The image conjures up chunks of metal sealed together—and that's where the metaphor falls short. When hormones bond with their chemical counterparts, they aren't so much locked but embracing in loose sort of way, like a couple dancing. They draw together; they drift apart; they reconnect; they drift apart; and sometimes a competing hormone breaks in and knocks the other one away from its dance partner. Though it seems as if the antibodies should be breaking in, as if they should be the entities chasing the hormonal invaders, in practice it is the hormones that knock one another off antibodies.

Berson and Yalow took advantage of this microscopic promiscuity to devise a technique they named radioimmunoassay, or RIA for short. Here's how it works. A scientist needs a known quantity of a hormone and a known quantity of the antibody, the immune cell that binds to it—the dance partners. Then, she adds to this mix the patient's blood, which contains an unknown amount of hormone. Now she has a known amount of the sample hormone, a known amount of antibody, and an unknown amount of the patient's own hormone.

Some of the patient's hormone will knock the original hormone off the antibody. Measuring the amount of hormone that is knocked off signals how much hormone is in the patient's blood. Though hormones are too small to measure directly, the hormone-antibody bond creates a larger chunk. And irradiating the sample hormone, so that it glows, makes it easier to spot. That's how Yalow and Berson could trace how much of the sample hormone fell off the antibody.

They devised a formula based on the tightness of the hormone-antibody bond (it varies among hormones). They plugged into their formula the measured amount of irradiated hormone that had been knocked off. A large quantity of knocked-off hormone meant there must be a large quantity of patient hormone to do the knocking off. Thus, they were able to quantify the amount of hormone in the patient's sample, down to one-billionth of a gram in a milliliter of blood.

Before RIA, if doctors wanted to estimate the potency of growth hormone therapy, for instance, they would inject a sample into a rat, wait two weeks for the stuff to kick in, and then measure the growth plate of the rodent's skinny leg bone. It was time-consuming and cumbersome. In comparison, RIA practically spit out the results.

With RIA, doctors could measure hormones for the first time. In the 1940s and 1950s, doctors had diagnosed patients with hormone deficiency without knowing how deficient they really were. They administered hormones without knowing how much the patient needed. When Jeff Balaban first saw Dr. Sobel in 1961, she did a lot of tests but she didn't measure his growth hormone levels. That was not yet possible.

Some colleagues suggested that Yalow and Berson patent RIA, but they preferred to make it widely available. "We had no time for such nonsense," said Yalow. "Patents are about keeping things away from people for the purpose of making money." So Yalow and Berson published the details of the test's inner workings in a 1960 article in the *Journal of Clinical Investigation*, and invited anyone who wanted to learn RIA to visit their lab, luring scientists from around the globe. Within a few years, RIA was a standard test used worldwide.

On April 11, 1972, a few days shy of his fifty-fourth birthday, Berson died of a heart attack while attending a medical conference in Atlantic City. Though she rarely showed emotion, Yalow sobbed at his funeral. She named their lab the Solomon A. Berson Research Laboratory so that his name would continue to be on all of her papers. She worried that her chances for the Nobel Prize were quashed without him, assuming that the scientific world considered him the brains and her the mere technician, because of her gender. She also assumed that no one would respect a lab run by a PhD rather than an MD. She was fifty-one at the time and considered going to medical school, not because she wanted to practice but to surmount potential barriers to the Nobel. In the end, she never got a medical degree, but devoted her time to working even harder in the lab, continuing to publish laudable research. In 1976, she was awarded the Albert Lasker Medical Research Award, considered a precursor to the Nobel. She got the Nobel the following year.

The history of endocrinology cannot be fully understood without knowing about RIA. RIA cannot be fully appreciated without knowing about Rosalyn Yalow, because her life is not just a story of a brilliant mind but of dedication and resilience. As the Nobel committee put it when they handed her the award on December 10, 1977, "We are witnessing the birth of a new era of endocrinology."

Yalow may have forged on, but she didn't forget the barriers along the way. By the time she got the Nobel, it was common knowledge that hormones elicit antibodies, that they provoke immune cells—just as she and Berson had proved in 1956, before anyone believed it. In her acceptance speech she brought up her original study, the one that no one wanted to publish. And she included her rejection letters in the Nobel exhibit of her work.

It's been said that from the moment of the Stockholm awards ceremony, she wore a Nobel Prize charm around her neck (given to her by her husband) and signed every piece of correspondence "Rosalyn Yalow, PhD, Nobel Laureate." It's also been said that Yalow tacked a sign on the bulletin board in her laboratory saying, "To be considered half as good as a man, a woman must work twice as hard and be twice as good." That's a common feminist maxim.

But Yalow added the punch line: "Fortunately, that is not difficult." Her children dismissed the jewelry/signature talk as the typical bluster of male colleagues. But they remember the sign well.

Yalow continued to lecture frequently, and she continued to conduct experiments until she was no longer able. In one of her last talks, she addressed a group of elementary school children in New York City, explaining how science often works: "Initially," she said, "new ideas are rejected. Later they become dogma, if you're right. And if you're really lucky, you can publish your rejections as part of your Nobel presentation."

In the mid 1990s, when she was in her seventies, she had the first of several strokes. She died on May 30, 2011, at the age of eighty-nine.

Soon radioimmunoassay had become a common item in a researcher's tool chest, just as stethoscopes are part of a clinician's uniform. By the 1970s, only a decade later, every endocrinologist had a tool to measure hormones down to the billionth of a gram. That's like being able to measure the extra water in a swimming pool after one swimmer sheds a tear. And not only could they measure hormones; they could distinguish one from its remarkably similar brethren. RIA transformed endocrinology from educated guesswork to a precise science. You could say the only immeasurable thing about RIA was its incalculable impact on medicine.

Dr. Thomas Foley was a young pediatric endocrinologist at the University of Pittsburgh and part of a team of doctors who decided to try RIA to detect hypothyroidism. He had heard about a pilot study in Quebec and decided to run a similar study. Foley still remembers the first baby whose test was positive, one of 3,577 babies tested. "It clearly improved our ability to determine hormone levels with relationship to disease. We didn't know a whole lot at the time, but it was pretty well established how beneficial this was," Foley recalled recently. Today, within moments of birth, pediatricians routinely take a dab of blood from a heel prick and screen newborns for hypothyroidism, so they can be given hormone treatment before any damage is done. An underactive thyroid gland is also caused by a lack of iodine, a mineral the body needs to make thyroid hormone; hence the global public health campaign to add iodine to salt. By the 1980s, both forms of "cretinism," congenital and acquired, were all but eradicated.

The detection of hypothyroidism was just a small part of RIA's impact. It was used to measure hormones for all kinds of suspected disorders. Today's fertility treatments wouldn't be possible without it. Beyond endocrinology, the tool has been used to measure other substances considered too tiny to quantify. Doctors could monitor drug levels and spot germs. RIA was used to detect HIV, the virus that causes AIDS. It's become so widespread that doctors cannot imagine how they ever worked without it. To be sure, today's RIA methods are not precisely the same as what Yalow and Berson created; even more sophisticated technologies have added tweaks to the original recipe. But the fundamental idea remains the same.

It would be easy to underestimate or ignore RIA altogether. It's technical, hard to understand. It's neither a cure nor a discovery; it's simply a way to measure. And yet it's hard to underestimate the significance of this invention, the impact it has had on the way science is done today. Radioimmunoassay provided doctors with a whole new vision. It was as if someone had lifted their blindfolds, and they could finally see what they were doing.

10.

Growing Pains

IN THE SPRING OF 1984, twenty-year-old Joey Rodriguez was on a flight from California to Maine to visit his grandparents. A few hours into the flight, he stood up and a whoosh of spins nearly knocked him over. His mother handed him candy. She figured his low blood sugar was acting up again. There seemed to be no reason to worry.

Joey had his share of medical issues. As a toddler he had been diagnosed with deficiencies in thyroid and growth hormone. His insulin system, which keeps blood sugar in balance, didn't work. He was given shots of all three hormones (thyroid, growth hormone, and insulin) throughout his teenage years. Instead of the usual three times a week, Joey had special permission from the National Pituitary Agency to get daily shots of growth hormone, because if he skipped one day of growth hormone, his insulin swung wildly. (Growth hormone affects not only growth, but sugar metabolism.) Sometimes, even with the right doses of every shot, his blood sugar plummeted and he'd suffer a dizzy episode. A little sugar did the trick, so his mother always had sweets with her.

During Joey's week in Maine, vertigo hit again. For days, he hadn't been feeling right. When his grandfather offered to take him for a motorboat ride, Joey replied that he "didn't need to go for a spin because he was already dizzy." At first his mother thought nothing of it, but as they were heading home, he got worse. It wasn't just vertigo: Joey seemed different; he wasn't himself. He stumbled getting off the plane. He walked as if he were having trouble balancing the weight of his own thin physique. He seemed drunk, but he wasn't. For the first time, talking was work. He spoke as if tiny weights were dragging his tongue to the floor of his mouth.

Ms. Rodriguez immediately took her son to Stanford University, to consult with the doctors who monitored his hormones. They couldn't find anything wrong. So she called the specialist who took care of him when he was younger. Dr. Raymond Hintz had started Joey on all of his medications and served as his primary caregiver for over a decade, until Joey aged out of the pediatric clinic. From a medical standpoint, Hintz knew Joey better than anyone.

When Dr. Hintz heard the fear in Ms. Rodriguez's voice—a woman he considered stoic—he told her to rush Joey to the emergency room. He met them there. Every imaging test, every brain scan, every blood test came back normal, so the hospital sent mother and son home. But Joey's mother wasn't ready to accept that her son, who was falling all over the place and slurring his speech, was okay. Every day, it seemed to her, he deteriorated.

She made an appointment with a neurologist. Joey clomped into his office, legs wide apart as if he would tip over if he held them closer together. He drooled. His shoulders slumped. His head wobbled back and forth. Enunciating words strained his jaw. Worst of all, he didn't seem to care one whit.

Utterly confused, the neurologist admitted Joey to the hospital, and then discussed his case at a weekly conference of specialists—from the dizziness on the plane to the cognitive decline. The doctors suggested a few possibilities, including an infection that he might have picked up in the Maine woods. But that didn't explain the airplane episode, which took place before he got to Maine. They wondered if he had inherited a degenerative disease, but they couldn't figure out what disease that might be. Dr. Michael Aminoff, a young faculty member—not yet a full professor—shot up his hand. "Creutzfeldt-Jakob disease," he said, referring to a rare, fatal brain disease, called CJD for short. Aminoff was working in the electroencephalogram lab, running brain scans. He had seen the electrical changes in Joey's brain and thought they mimicked those he'd seen in adult CJD victims. He also said that Joey seemed like a CJD patient—suffering from rapidly progressing dementia without any other cause.

The senior doctors dismissed his suggestion. First of all, young people then didn't get CJD; the typical patient was about eighty years old. Secondly, CJD doesn't start with clumsiness. It starts with dementia.

CJD cannot be picked up by any test. The EEG, a kind of brain scan, is a clue, but not a definite diagnosis. The only way of telling whether a patient has it is by examining the brain on autopsy. A pathologist can easily spot the telltale sign of the disease: a spongy, holey brain.

When Aminoff—who is now a neurologist and director of the University of California San Francisco's Parkinson's Disease and Movement Disorders Clinic—read about Joey, he wondered whether the growth hormone he'd been given was contaminated. "I said they should go back and inquire about the donors"—the cadavers whose pituitaries had been mined for growth hormone—to see if they had had a brain disease. The senior doctors paid no attention to that either. His words were dismissed as the naive conjecture of an overly enthusiastic, inexperienced youngster.

Six months after the plane flight, Joey died, having never seen his twenty-first birthday. The autopsy showed that his brain was spongy and holey. Clearly he had died of CJD. Within a few years, Joey's CJD—and that of hundreds of children like him—would be tied to contaminated growth hormone.

The story of growth hormone is an example of everything that can go right and everything that can go wrong with a medical discovery. It combines the ingenuity of scientists, the hubris of doctors, and the desperate commitment of parents. The biggest fear was that it wouldn't work, that it wouldn't make a child grow. The tragic reality of contamination did not emerge for years.

In the beginning, everyone was on board. The parents of the 1960s had been children in the 1940s, when antibiotics came on the scene, which were ballyhooed as a way to wipe out infectious diseases once and for all. They were teenagers in the 1950s, when folks were lining up for the polio vaccine, which was touted to eradicate this crippling menace from the planet. They were not the skeptics we are today, wary of hidden toxins. They believed in medical science. They believed in all the good it had to offer.

And they were activists. They marched against war, for civil rights, against segregation. They had a we-can-do-it mindset; they demanded medicines they deemed their right. They were worried but optimistic, desperate yet organized. The same optimism that energized Barbara Balaban to collect pituitaries blinded her to the potential downsides.

The growth hormone saga is also a story of the clinicians, who were just as excited by the hormone headlines, just as eager as the parents to help these young patients feel a little more normal. They were the ones on the front lines doling out the vaccines and antibiotics, so they were equally gung-ho about all that medicine had to offer, maybe even more so. Many of the old guard had watched the rate of death in childbirth plummet, and were seeing patients live longer than ever before thanks to the wonders of modern medicine.

There is more to the story than naïve parents and audacious doctors. As one endocrinologist said years later, everything is easy through a retrospectoscope. In other words, it's always easy afterward to see a clear path to the culprit, but in the fog of the journey, clues—and even warnings—are often seen as random, inconsequential weeds in the forest.

When Joey's CJD was revealed after his death, Ray Hintz, his pediatrician, panicked. CJD is rare. It strikes about one in a million people every year. There are many CJD-like brain illnesses: mad cow disease in British cattle, scrapie in sheep, and kuru in a tribe in Papua New Guinea, to name a few. Doctors lump them all into one category called transmissible spongiform encephalopathies, or TSEs. The name itself reveals a lot about what we know and what we don't know: the disease can be transmitted; it creates spongelike holes; it targets the brain.

Hintz recalled someone saying something about the possibility of infected brain tissue getting into growth hormone medicine at a hormone conference two years earlier. At the time, it seemed a highly unlikely, hypothetical situation. Now it seemed like reality.

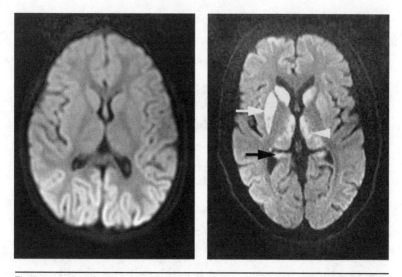

The brain on the left is an image of a normal healthy brain; the brain on the right is a CJD-infected brain from cadaver-sourced growth hormone. *Left image courtesy of Dr. William P. Dillon, University of California, San Francisco; right image courtesy of Peter Rudge, MRC Prion Unit at University College London.*

On February 25, 1985, Hintz wrote a letter expressing his fears to the U.S. Food and Drug Administration, the National Institutes of Health, and the National Pituitary Agency. The NIH administrators, in turn, telephoned pediatric endocrinologists, urging them to check on their former growth hormone patients. They needed to find out whether Hintz had picked up a fluke or uncovered a connection.

On March 8, 1985, a group of growth hormone experts met in Washington, D.C. Most of them were skeptical. Many were angry. After all, they were talking about only one boy. They worried more about a national panic than a national epidemic. If fear spread needlessly, thousands of children could be deprived of crucial treatment—all because of one random death.

Dr. Robert Blizzard, who spearheaded the collection of pituitaries, remembers thinking that Hintz, a close friend of his, was reacting too quickly. One case doesn't mean a trend, he said.

Blizzard, for his part, had given himself shots of growth hormone. When he was treating children with growth issues, the thing that struck him, besides their lack of height, was that many of them looked old. Their skin was wrinkled; their faces had lost cheek fat. Blizzard had wondered whether a lack of growth hormone made them age too fast. Then he had wondered whether shots of growth hormone could slow the aging process. Or better yet, could growth hormone reverse the clock? Unwrinkle your face? Restore hair color? In 1982—a few years before Ray Hintz sounded the alarm—he'd tried it on himself and corralled a few friends to do the same. They took one milligram every day. "I was on it for the full two and a half years, the other fellas for a year and a half," Blizzard told me.

Blizzard monitored key metabolic indicators and measured bone density. He even studied the men's fingernails. "I never put this into the press," said Blizzard, "but I learned what I wanted to learn, which is that it didn't make your hair turn from gray to black and the girls didn't whistle at you."

But the possibility of growth hormone killing children? Nonsense.

Carol Hintz, Ray Hintz's widow, remembered those days well. (Ray Hintz died in 2014.) "It was a very difficult time," she recalled. "Some of the endocrinologists were perturbed and thought he was stirring the pot. They just couldn't believe it. Doctors would call him at home and say, 'What do you think you're doing? There's nothing wrong here.' Dr. Blizzard had used it on himself and he was fine and still going strong. Other people tried to say Joey was on drugs or something like that. My husband knew the family well and said it just wasn't possible."

One month after the experts convened, and one month after Blizzard pooh-poohed the notion of dangerous growth hormone, Blizzard got a call from a doctor about one of his former patients. A thirty-two-year-old man from Dallas, Texas, had died the same way Joey Rodriguez had: drunken gait and rapid descent into dementia. He, too, had been on growth hormone for years. His doctors had assumed he had a motor nerve disease, perhaps multiple sclerosis.

Then Dr. Margaret MacGillivray, a pediatric endocrinologist, got a call from the family of a former patient, a twenty-two-year-old man from Buffalo, New York. Same thing: loss of motor control, then senility, then death. No one had connected his illness to growth hormone either. No one had thought to call his former pediatric endocrinologist when neurologic symptoms arose.

Three cases turned Dr. Blizzard's nonchalance into concern. Or, as neuroscientist Paul Brown would write in an article about the history of growth hormone, "The effect of the new information was like two thunderclaps and forever sealed the fate of native growth hormone therapy."

The growth hormone experts convened again on April 19, 1985. This time, no one was calling Dr. Hintz an alarmist. The Food and Drug Administration banned almost all human-growth-hormone therapy. It was permitted only for children with such severe hormone deficiencies that they would die without it.

Shortly thereafter, the FDA green-lit a version made in the lab by Genentech, rocketing the company from a small startup into a major biotechnology company. As Dr. Brown noted dryly, "Only Genentech is not in mourning." Until the growth hormone fiasco, hormones from humans or animals were considered natural and therefore safer. Fears lingered about laboratory-made versions. The balance shifted after the deaths began. All of a sudden, synthetics seemed purer, less toxic. The ban didn't stop growth hormone therapy completely; it merely switched one kind (pituitary-derived) to another kind (lab-made).

Doctors, like everyone else, are swayed by politics, by large-scale fears, by the culture of the time. Human growth hormone, the kind from cadavers, was distributed in the 1960s and 1970s before widespread concern that contagion might lurk in body parts. To be sure, tissues were tested for a short list of known viruses, but there was not a huge emphasis on preventing unknown diseases. The thinking was, as one biochemist said, that the product came from human tissue, and how could human tissue harm other humans? The tragic reality that batches of growth hormone had transmitted a deadly agent came to light during the AIDS epidemic in the mid-1980s. Suddenly, the notion of lurking diseases made sense.

In the meantime, the NIH started the unwieldy process of reaching out to every single growth hormone recipient, all 7,700 of them. It wasn't easy, because patient privacy meant that names of growth hormone patients were not kept in a registry. Officials sleuthed through data banks where patient names had been replaced with codes, tracking down doctors who might remember patients from years past. Some doctors had retired. Some records had been thrown away.

Finding the patients was not the biggest challenge. When growth hormone was extracted from human pituitaries, laboratories had pooled it into huge batches. There was no way to tell who got hormone from which gland. Even if NIH officials could identify the tainted batches, no one knew if the diseased hormone had been mixed with the clean stuff.

Hundreds of hormone recipients who had once considered themselves the lucky ones, because they got a portion of the limited supply, now saw themselves as potentially doomed. The Mahlers, Balabans, and thousands of others received a two-page letter from the National Institute of Diabetes and Digestive and Kidney Diseases, dated November 27, 1987. It said, in part, that some of the growth hormone their kids had received years ago might have been tainted with a deadly disease. The letter warned parents not to let their children donate blood because they might pass along a deadly agent. What parents really wanted to know was whether their kids already harbored it.

No one could say. The disease agent can lurk in the brain for decades before igniting a physical and then a cognitive decline. Once activated, it kills quickly—often within six months after symptoms first appear. No one knew whether the five deaths were a fluke, the end of tiny yet tragic episode, or the beginning of an epidemic. Only time would tell.

The Balabans received the letter when Jeff was thirty-five and living in California. "I don't think I told Jeff right away," Barbara Balaban said. "I think we worked out what we would say. We were careful how we presented it, in terms of other kids had a bad reaction." They don't remember saying the words "deadly brain disease."

Larry Samuel, a lawyer in New Orleans, was given growth hormone shots too. He said he wasn't "panicked or angry, but I had questions and Bob [Blizzard] was always frank with me and he was concerned. I want to say that, okay, oh gee, maybe about five years ago, after Katrina—that's how we base our lives down here—I developed a tremor and it was quickly diagnosed as non-Parkinson's. I called him [Dr. Blizzard] and said, 'Is this something,' something related to CJD?"

David Davis is a journalist who received growth hormone shots. He interviewed other recipients and wrote that "the overwhelming feeling I got from my interviews with others was one of abandonment: the people who got us into this mess have completely abandoned us. Once a year—tops—they send us an update."

The news of the tainted hormone prompted other countries to wonder whether this was an American problem or a worldwide one. Sure enough, when they looked, they found similar deaths. Sarah Lay, a young woman in England who had been on growth hormone, died of CJD in 1988. Others would emerge. British officials decided, at first, not to alert patients as they didn't want to terrorize the public.

Then there was a death in Australia. Australia decided to contact doctors and leave it up to them whether to pass on the news.

Soon enough, nearly the entire world shuttered the cadaver-derived hormone businesses. The U.K., New Zealand, Hong Kong, Belgium, Finland, Greece, Sweden, Hungary, West Germany, Argentina, and the Netherlands closed shop on the natural hormone. But not France. Dr. Jean-Claude Job, the pediatrician who headed France Hypophyse, the French pituitary agency, decided to add an extra purification step rather than switch to the laboratory version. He would not halt human growth hormone production for another three years, a delay that would come back to haunt him.

There had been naysayers from the beginning. Dr. Alan Dickinson, the director of the Neuropathogenesis Unit in Edinburgh, was an expert on scrapie, the sheep version of CJD, which had been around for years. In 1976, he wrote a letter to the U.K. Medical

Research Council warning that pituitaries could be tainted with CJD. No one cared, he said.

Another naysayer was Dr. Albert Parlow, who ran a lab at Harbor-UCLA Medical Center in Torrance, California, that processed pituitary hormone. Around the same time that Dr. Dickinson was sounding the alarm, Dr. Parlow had expressed concern that the processes being used in other U.S. facilities to extract the pituitary hormone did not include sufficient purification. Some believed that including additional purification steps would result in a lower yield of the hormone, which was always a concern given that there were so few sources of the hormone to start with. But Parlow's method included an additional purification step, which, Parlow believed, yielded a purer product.

A study published in 2011 of 5,570 people who had been treated between 1963 and 1985 appeared to confirm Parlow's fears. In 1977, the National Pituitary Agency had switched all of the pituitary processing to Dr. Parlow's lab, not out of safety concerns, but because Parlow's extraction method actually gleaned more hormone from each pituitary than other processes: 7 milligrams versus the standard 1 milligram. The 2011 study found that all 22 CJD victims in the U.S. had gotten their hormones before Parlow's laboratory took over the processing. The team of investigators, who included researchers from the Centers for Disease Control and Prevention and the National Institutes of Health, concluded that Parlow's purification method "greatly reduced or eliminated" the CJD agent. Dr. Salvatore Raiti, who was the director of the National Pituitary Agency, said he had no doubt that "[t]here have been no cases from the later hormone because we had better extraction techniques and we had better knowledge."

To this day, the NIH continues to track U.S. growth hormone

recipients. Since 1985, when tracking began, there have been 33 confirmed deaths among 7,700 Americans treated. In France, the total has reached 119 deaths, among 1,700 people treated (as many as in all other countries combined, and by far the worst ratio). In the U.K., there have been 78 deaths, plus one person diagnosed with CJD in August 2017 and still alive, among 1,849 people treated; in New Zealand, 6 deaths among 159 people treated. Holland and Brazil reported two cases each. Austria, Qatar, and Ireland have each reported one death. All were ascribed to CJD.

A few American families tried to sue their doctors or the NIH, but no individuals or organization were found guilty of negligence or malpractice. For the most part, the courts found that doctors were practicing standard-of-care medicine. Most claims didn't even make it that far.

In 1996, British courts ruled in favor of the patients, putting aside $7.5 million to compensate not just the families of those who died, but anyone who might have been treated with potentially contaminated growth hormone.

In 2008, a group of French families sued seven doctors and a pharmaceutical company on charges of manslaughter and deception. They lost the case. "I fear we may have not learned any lessons from this case and will face other and bigger public health scandals in the absence of adequate scientific and medical caution over the effects of new treatments on young people and future generations," announced Dr. Luc Montagnier, who won the Nobel Prize for isolating HIV, the virus that causes AIDS. He had served as an expert witness for the families.

It's easy to blame the entire medical profession for the growth hormone tragedy. But many doctors, Blizzard included, believe that more good than harm emerged from the science. Jeff Balaban was one of the lucky ones, along with Larry Samuel; they both grew a few inches thanks to the medicine, and never suffered toxic side effects. If there is one hero in the story, it is Ray Hintz, the doctor who made the seemingly unlikely connection. When his patient, Joey Rodriguez, died of a rare brain disease, the doctor could have assumed it was bad luck, some inborn mutation, a rare infection he picked up somewhere. But Hintz had two things going for him: he remembered the comment in a meeting years before, and most importantly, he knew Joey and his family. When Joey got sick, he was there by his side, listening and watching, picking up the crucial clues, the kind that don't depend on lab tests but on doctors taking in what their patients are telling them. Hintz rightly sounded the alarm that would uncover a mystery, one that could easily have gone undetected for years.

11.

Hotheads: The Mysteries of Menopause

FLORENCE HASELTINE, an obstetrician–gynecologist, was about as entrenched in women's health as anyone could be. She founded the Society for Women's Health Research and served on the board of American Women in Science. She had been a director of the National Institutes of Health's Center for Population Research and an associate professor at Yale University. In addition to her medical degree, she earned a PhD in biophysics from the Massachusetts Institute of Technology. She also coauthored *Menopause: Evaluation, Treatment and Health Concerns*, a summary of the most up-to-date information. Haseltine was an insider. She had access to behind-closed-doors talk among medical experts.

And yet, when she noticed the first hints of her own menopause, she opted for a medical treatment that appalled her colleagues. There's not even a mention in her own book of the course of treatment she prescribed for herself.

In the summer of 1990, when Florence Haseltine was forty-eight years old, she convinced a gynecologist to give her a hyster-ectomy, an operation to remove her womb. There was no pressing medical reason. She had neither a painful growth nor a cancer, the usual reasons for the operation.

When Haseltine made her decision, she was no longer on staff at Yale, but commuted weekly between her husband and daughters in New Haven and her job at the National Institutes of Health in Bethesda, Maryland. She didn't want the hysterectomy to be done at Yale, because she knew it would rankle her former colleagues. She was never one to avoid an argument, but she didn't want her personal decision to become fodder for debate and gossip. So she went back to the hospital where she had done some of her medical training. "I called my favorite gynecologist in Boston and said, 'Sign me up before Labor Day.'"

Haseltine wanted to take estrogen to quash her hot flashes—flare-ups of drenching sweat, intense heat, and flushing. But estro-gen, she knew, increased the risk of endometrial cancer, in the lining of the uterus. That's why she wanted her womb out. Sans uterus, she could take the hormone without worrying.

"I had terrible hot flashes, even with my menstrual periods," she explained years after the surgery. "I just looked at all the data on hormones in the 1980s and it was all there."

Haseltine was well aware that she could have skipped the sur-gery and added progesterone to the estrogen regime. Progesterone, she knew, counteracted the increased risk of uterine cancer. But she didn't want to take progesterone. "It makes you feel shitty and increases bleeding," she said. "There's not a word in the English language to describe how miserable it makes you feel. That's why I had the hysterectomy—because I wanted estrogen and not proges-terone and it also eliminated the risk of cervical cancer." With the womb removed, there's no way to get cervical cancer, cancer of the opening of the uterus. Cervical cancer risk is increased by HPV, a sexually transmitted virus. Or as she put it, "I'm a child of the sixties and you expose yourself to a lot of partners, so we eliminated two for one."

Haseltine has been on one milligram of estrogen a day ever since.

Around the time that Haseltine was trying to minimize the symptoms of menopause, Helen E. Fisher, an anthropologist at the American Museum of Natural History, wrote an article extolling the wonders of middle-age hormonal swings. She contended, in a 1992 opinion piece in the *New York Times*, that the lower estrogen and slightly higher testosterone levels of menopausal women makes them more assertive and aggressive in the workforce. "And the biological changes wrought by menopause will bolster their interest in power and increase their ability to use it."

Maybe so. Maybe the renewed confidence would allow "Boomer women to attain authoritative political positions," as Fisher put it. She didn't cite any science in the editorial to back up her claims. It seemed more like a way to make women feel better about their aging bodies and age-old workplace situation. Or maybe it was a way to tell the world that menopausal women had a lot to offer at the office and shouldn't be put out to pasture when their fertile years ended.

Despite Fisher's sanguine prose, menopause made a lot of women feel lousy. The end of periods, as generations of women already knew, can mimic the beginning of them. Many women are slammed by bouts of inner rage, the kind they haven't experienced since teenagehood. Inner dialogue becomes a rant of snarky one-liners that can accidentally turn into outer dialogue.

Then there are the hot flashes. The name is a misnomer. "Hot flash" sounds like a flicker: short and quick, no big deal. But it's more like an abdominal furnace on full blast creating a suffocating swelter. For most women—about 80 percent—hot flashes strike when they reach their fifties and weather on for a few years, sometimes during the day, sometimes making for sleepless nights. The British call them hot flushes, which sounds more like a toilet metaphor, but also like a slow swirl, which is more like the way they really are.

For an unfortunate few, the symptoms linger for decades. For a lucky minority, they never happen. Some women skip through the whole experience: their periods stop and that's that. No erratic temperature changes; no mood swings; no brain fog; libido as good as ever. To those women, the rest of us must seem like cranky bitches.

Haseltine had her surgery in the 1990s, a decade that, as she put it, ushered in a "sea-change of interest in menopause." Many of the women who demanded information about menopause were the same ones who had pushed for safer versions of the birth control pill years earlier. Their concerns aged along with them. As they reached the end of their childbearing years, the focus of their activism shifted from contraceptive hormones to menopause hormones. Menopause and its accompanying issues made front-page headlines, garnered the leading spot on the evening news, and even slipped into a few sitcoms. It's not that no one talked about menopause before the 1990s, but the level of discourse took on a new urgency. Women's issues had come to the fore, in part thanks to Bernadine Healy, the first female director of the National Institutes of Health, who was appointed in 1991. Under her leadership, funding for women's health research increased.

A few NIH studies suggested that taking hormones after menopause not only alleviated symptoms but also warded off diseases of old age, such as Alzheimer's and heart disease. Doctors and pharmaceutical companies were excited about the supposed benefits. Yet despite the enthusiasm, aging women were confused. They wanted to know two things: 1) what to do about menopause, and 2) what was going on inside their aging bodies. Clues were beginning to emerge.

Dr. Robert Freedman, a professor of psychiatry and obstetrics and gynecology at Wayne State University, is a leading investigator of hot flashes. His initial research had nothing to do with menopause. In 1984, he was studying whether biofeedback (using thoughts to alter physical symptoms) helped people with Raynaud's disease, which causes painfully cold hands and feet in chilly weather. "One Friday afternoon," Freedman recalled, "a graduate student came into my office and he said, 'I've read your studies and found you can take cold women and make them warm. Can you take hot women and cool them off?'"

The student's mother suffered from hot flashes. Freedman hadn't given much thought to menopause, but was intrigued by the challenge. So he advertised for volunteers in the local newspaper. He hoped for just a few but was flooded by responses, women anxious to try anything to help them get a good night's sleep, to prevent the torrents of sweat.

Freedman developed a method to ignite hot flashes in the laboratory and another technique to monitor them objectively. During each session, the woman would lie on a recliner in a room that got progressively hotter. She was also wrapped in pads that contained heated water, sort of like an electric blanket, or, as Freedman explained, like the cover used to warm newborns or laboratory animals. In order to know precisely whether the woman was having a hot flash, he attached EKG-like leads to her chest. The leads recorded electrical conductivity. The salt in sweat bumps up conductance. That signaled the start of the heat wave. Lastly, to gain information about the woman's core body temperature, he used a digestible thermometer, the size of a large pill. The women swallowed it, like an aspirin. The tablet transmitted temperatures every thirty seconds to a receiver worn on a belt or located elsewhere in the lab, as it traveled from mouth to anus. "It traverses your gut for whatever your transit time is," explained Freedman, "and is then excreted with feces. They don't have to be retrieved. Although my excellent, late chief engineer, Sam Wasson did retrieve the first one to dissect it and determine how it worked."

Freedman tried all sorts of techniques to see which one, if any, would minimize hot flashes. The most effective method, he said, was deep, abdominal breathing for fifteen minutes twice a day. It alleviated daytime hot flashes, but not nighttime ones. "It's problematic at night," he said. "We haven't figured out a way to implement that in a nighttime schedule."

Most people don't notice a small fluctuation, about half a degree Fahrenheit, in core body temperature—the heat deep inside the body. A greater, downward change in temperature makes one shiver to get hotter. An upward change makes one sweat to get cooler. In menopausal women, that narrow window of climate control slams shut. A tiny uptick in core body temperature can spark a tsunami of sweat. That's why a menopausal woman may be fanning herself in a mildly warm room when no one else seems bothered. That's why, when the temperature rises a tad during the night, a menopausal woman is flinging off the blankets and pillows while everyone else is snoozing comfortably (everyone except the bedmate attacked by hurling covers).

Because hot flashes occur when estrogen plummets, scientists have long assumed that the two are related—though it's not the level of estrogen that matters, researchers have discovered, but the drop. Women with chronically low levels of estrogen do not suffer hot flashes, but researchers can trigger a hot flash if they put low-hormone women on estrogen and then take it away. They have also noticed that during a hot flash, adrenaline, the fight-or-flight hormone, increases. That may be why some menopausal women say they feel panicked, particularly in hot, closed spaces—an awful anxiety that they may never have experienced before menopause.

But while these various physiological events have been documented (estrogen plummets, adrenaline rises, blood vessels dilate), it is not yet known how they are connected. Does the drop in estrogen level cause the increase in adrenaline, or is another hormone involved?

The way people react to changing temperature is complicated. A crisscrossing network of nerves and hormones connects temperature receptors in the skin to deep organs. When temperatures fluctuate and alter the body, scientists can observe the aftermath. But it's tricky to figure out the order of events. It's like finding a cobweb and trying to decipher how it all came together.

A major impediment to research is that there aren't any good animal models. Humans seem to be the only creatures that get hot flashes. "I spent four years of my life trying to make rhesus monkeys flash," said Freedman. "We took out their ovaries. We took out estrogens. We heated them up. Nothing worked."

Some scientists say killer whales have hot flashes—making them the only menopausal mammal other than humans. The evidence is tantalizing. Female killer whales live many years after they stop making babies, prompting scientists to posit that they experience menopause. They start making babies when they are about twelve years old and stop in their late thirties or early forties. Yet they live into their eighties. That suggests they may experience the same hormonal changes as humans. Even if the theory pans out, it's no use to Freedman. He needed volunteers in his lab, something more manageable than menopausal killer whales.

While Freedman was making women hot, Dr. Naomi Rance, a professor of pathology at the University of Arizona, was digging deep into the cellular aspects of menopause, examining dead women's brains. In the 1980s, Dr. Rance had completed her medical training and was finishing her PhD in neuropathology at Johns Hopkins University, exploring the hormonal changes of puberty. But as she aged, so did her interests. She switched from exploring puberty to exploring menopause.

Collecting brains from dead women wasn't easy. Rance needed ones that weren't riddled with disease, such as Alzheimer's or cancer, which would clutter the study with too many variables. And she didn't want to rely on other pathologists, because she needed to make sure the organs were removed carefully, without tugging at the parts she needed to examine. She trusted her own techniques.

"I took the brains out myself because, as a neuropathologist, part of what you do at autopsy is take out brains and slice them and look at them and find out what's wrong with them and why they died." She needed the hypothalamus, which contains hormones that control reproduction. It's at the base of the brain. She also needed the pituitary, which hangs off the brain and also contains hormones that control reproduction. "You have to be careful not to rip the stalk of the brain stem," she said. Plus, she needed the brain to be fresh. "I had to aim for less than sixteen hours postmortem. My limit was twenty-four hours." A longer delay might alter the cells that she wanted to examine.

For her first study, she collected three brains from younger women and compared them to three brains from older women. It was a small study, but the differences were astonishing. Rance found that one particular brain cell, a neuron in the hypothalamus, was 30 percent larger in the older women than it was in the younger women. The difference, she said, was "like night and day." In an image in her article published in the July 1990 issue of the *Journal of Clinical Endocrinology and Metabolism*, the hypothalamic neurons of the postmenopausal women are about the size of blueberries. Those of the premenopausal women are the size of capers.

Rance examined the same feedback loop—the ebb and flow of the hormone system—that had led to the development of the birth control pill. She suspected that in menopause, the brain gets the signal that estrogen is low and fires up cells that should lead to an estrogen boost. But because the ovaries are no longer functioning, the estrogen level doesn't rise. The brain keeps getting bombarded with we-need-more-estrogen messages. The continued assault pumps up the cells.

To test her theory, she studied six more brains, from three premenopausal women and three postmenopausal women. This time, she found that one particular type of cell was swollen in the older women, and found an abundance of estrogen receptors. She also zeroed in one on chemical, neurokinin-B, that might be responsible for some of the changes in menopausal brains.

A British team, more recently, found that injections of neurokinin-B sparked hot flashes in women—a clue, but not conclusive. The current theory is that the swollen cells in the hypothalamus are probably messing up older women's internal climate control system. It's not a complete picture, but the beginning of a sketch. Based on these findings, doctors have recently begun to test a drug that blocks neurokinin-B as a nonhormonal way to cool hot flashes. Preliminary results are promising.

By the 1990s, when Rance was digging into the nitty-gritty of brain research, she was hoping her insights would lead to better treatments for menopausal women. At the same time, another group of hormone researchers was considering the same issues, but from a much broader perspective. Rance was taking the view from the inside out, deep inside the brain cells. These researchers were looking from the outside in. They weren't thinking about cells and their proteins, but about people and the risk of diseases. They noticed, for instance, that compared to younger women, older women were more likely to get heart attacks, Alzheimer's disease, osteoporosis (bone thinning), and certain kinds of cancer. They also noted that older women have less estrogen. Could the two be connected in some kind of cause-and-effect relationship? In other words, does estrogen protect younger women from these diseases? And if so, could older women protect themselves from these same diseases by taking estrogen?

Notions such as these changed the way menopause was considered: it began to be seen as more akin to a hormone deficiency, such as diabetes, than as a natural part of aging. A series of studies about the value of hormone replacement therapy followed. For anyone who has tried to keep up with the news reports, it seems like a series of flip-flops.

Hormone replacement therapy is good for you; it's bad for you; take the drugs for a few years; take the drugs forever. For the most part, the majority of women taking the drugs were white and upper-class. A 1997 study that examined national statistics from 1970 to 1992 found that black women were 60 percent less likely than white women to use hormone replacement therapy. Another study, which gathered data from more than 30,000 office visits in the first half of the 1990s, found that while hormone replacement therapy prescriptions increased overall, white women were twice as likely as black women to get a prescription, and patients with private insurance were nearly eight times more likely than Medicaid patients to get the drugs. Was it because the pharmaceutical companies targeted white, upper-class women? Or were these women more likely to ask their doctors for a remedy for menopause?

Looking back on it all, a group of historians and scientists who met for a two-day conference in 2004 wondered whether history would have been shaped differently if, instead of calling the drugs hormone therapy, they'd been called hormone manipulation. Perhaps. But it is a rare salesperson who labels their wares manipulators.

In the 1910s and 1920s, extracts from cow and sheep ovaries were used to treat women bothered by the hot flashes and headaches of menopause. McElree's Wine of Cardui, taken three times a day, was touted to help with menstrual irregularities and "change of life," the name for menopause. It contained 20 percent alcohol. Beginning in the 1940s and 1950s, women were given the purified substance, pure estrogen. The pills became tremendously popular after the 1968 publication of *Feminine Forever*, a book by Dr. Robert Wilson that advised women to take estrogen not just to alleviate symptoms, but to maintain their youthful glow. "No woman can escape the horror of this living decay," he wrote. But there was a solution. "If a woman's body is furnished through pills with the needed estrogen (no longer supplied by her ovaries), her rapid physical decline in post-menopausal years is halted. Her body maintains its relative youthfulness just as a man's does." What Wilson didn't mention in his book was that he had founded an organization, the Wilson Foundation, which was funded by three drug companies:

McElree's Wine of Cardui for menstrual irregularities and change of life (an early twentieth-century term for menopause). *Division of Medicine & Science, National Museum of American History, Smithsonian Institute.*

Searle, maker of Enovid, the birth control pill; Ayerst, maker of Premarin, an estrogen pill; and Upjohn, maker of Provera, a progestin, a synthetic form of progesterone. His book was sold as an expert's advice, but it was really one large advertisement.

In many ways the menopause-hormone story is an extension of the birth control pill story. The hormones are the same—a mix of estrogen and progesterone. For menopause, they're dubbed "hormone replacement therapy," or HRT. The contraceptive pill and HRT were both hailed as a triumph for women's health, and then feared because of toxic side effects.. In both cases, the decision tree was tricky because there was no real illness: the drugs neither cured a disease nor prevented one. Women were taking pills to help them navigate two crucial times in their lives: to prevent an unwanted pregnancy and to prevent unwanted menopausal symptoms.

The birth control pill, approved by the U.S. Food and Drug Administration in 1960, was the first drug prescribed to healthy people for social reasons—a drug that didn't even promote well-being. It's the only pill called "the pill." Scientists got the idea from something that farmers had observed for centuries: you can't get pregnant when you're pregnant. So they created a therapy that mimicked some of hormonal changes of pregnancy. By the 1970s, enthusiasm had waned; the birth control pill was linked to deadly strokes and heart attacks, and also to nasty side effects, including depression and bloating. These findings, broadcast by women's health activists, prompted drug companies to create a lower-dose pill and pushed the government to mandate package inserts that listed risks.

At the same time, in the 1970s, scientists discovered the link between estrogen taken for menopause and uterine cancer. Women were flummoxed. They had heard about nothing but the benefits of hormone therapy to keep their bodies balanced, and now it seemed they had been poisoning themselves. Prescriptions for hormone replacement therapy nearly halved, from 28 million in 1975 to 15 million by the decade's end. Soon thereafter, researchers figured out that adding progesterone to estrogen counteracted the risk of getting cancer of the uterus. Sales crept back up.

By the time Haseltine started taking hormones, estrogen (alone or in combination with progesterone) had begun to regain popularity. Without much evidence, but based on theories and clues, a consensus emerged that estrogen warded off diseases of old age. The message switched from hormones for sex appeal to hormones for wellness. One study, called PEPI, found that women who took estrogen had better markers of a healthy heart, such as lower cholesterol levels. Another massive study, which tracked the health of more than 100,000 nurses, suggested that those who took estrogen had lower rates of heart disease. In 1992, the American College of Physicians advised that every woman consider long-term hormone therapy in order to reduce the risks of heart attack (the number one killer of women) and Alzheimer's disease (the number one fear). Shortly thereafter, other studies suggested that hormone therapy (estrogen or estrogen-progesterone combined) lowered the risk of colon cancer. There was some bad news (one study linked estrogen to breast cancer) but it was buried among all the good news. For a time in the 1990s, women were asking for hormone therapy because they thought it was wise for the long haul—not simply to quash the nagging symptoms. The number of prescriptions more than doubled, from 36.5 million in 1992 to 89.6 million in 1999. Hormone replacement therapy was the most popular drug in America.

Still, questions lingered. Quite a few doctors noticed the dearth of data. And so it was that a group of experts initiated one of the largest studies ever done on the long-term effects of hormone therapy. It was called the Women's Health Initiative. From 1993 to 1998, more than 27,000 women were randomized to receive hormones (estrogen alone if they had had a hysterectomy, estrogen plus progestin, a synthetic progesterone, if they hadn't) or a placebo. In the beginning, some doctors were so confident about the benefits of hormones that they believed withholding them from the women in the placebo arm was unethical.

Interim data showed just the opposite. In 1998, another,

smaller hormone study led to the shocking conclusion that women who already had heart disease and who took hormones had an increased risk of heart attack shortly after starting treatment. However, this was just one study, and everyone was holding out for the results of the larger Women's Health Initiative study of generally healthy women. In July 2002, the WHI estrogen-plus-progestin trial was halted abruptly, three years before it was supposed to end, because researchers had detected that the women taking hormones had more strokes, blood clots, and breast cancer than the women who weren't taking hormones. The headlines shocked, scared, and enraged women. They glanced at the headlines and assumed that all hormone therapy for menopause was dangerous—that it didn't work at all, that it was dangerous for all women.

But the study was not about taking estrogen and progesterone to ameliorate the symptoms of menopause; it was not limited to women who had recently hit menopause and might be taking hormones for a few years. It was designed to investigate the effects of the hormones on women long past menopause. The average age of the subjects was sixty-three. "The goal of the WHI was completely different from how people interpreted it," said Dr. JoAnn Manson, a Harvard University professor of medicine and one of the investigators. "The goal was to evaluate the balance of benefits and risks of hormone therapy when used for prevention of heart disease and other chronic illnesses. It was not designed to evaluate whether hormone therapy was safe and effective for the short-term management of symptoms. Any extrapolation of the findings to women in their forties and fifties is inappropriate." Also, as Mary Jane Minkin, a menopause expert and an obstetrician-gynecologist at Yale University, pointed out, the WHI used Provera (or progestin), a synthetic progesterone that was popular in the 1990s. Nowadays, many more doctors tend to prescribe Prometrium, a natural form of progesterone that has been shown in some studies not to be linked to an increased risk of breast cancer. The WHI was also limited to pills; there are other choices of hormone therapy, such as patches and gels.

The bottom line, as Manson explained, is that, contrary to earlier hopes that hormone therapy would prevent diseases of old age, it doesn't. So, hormones should not be used for disease prevention. They should be used for symptoms of menopause. Still, the news frightened women so much that the demand for hormone replacement therapy tumbled. Prescriptions fell by nearly half for women taking the combination estrogen–progesterone pill and by nearly one-fifth for women taking estrogen alone.

The latest WHI results, released in September 2017, found that after eighteen years there was no difference in death rates between the group of women taking hormones versus the group that didn't. Manson told Reuters that these results should reassure women who were worried about an increased risk of stroke, breast cancer, or heart attacks.

For women who opt for hormones, though, the choices today are befuddling. There are pills and patches and intrauterine devices that squirt hormones. Estrogen and progesterone come in a variety of doses. There are also so called compounded hormones, which means they are custom-made for each patient. Compounded hormones are good for the rare person who may be allergic to an ingredient in a pill (such as peanut oil) or for the individual who can't swallow pills. But the way they are advertised, you'd think you are getting the free-range, grass-fed equivalent of hormones—pills made just for you in small mom-and-pop shops.

Buyer beware. Many of the compounded drugs are made in factories, just like Big Pharma pills. Since the 1990s, the custom-made hormone therapy business has skyrocketed into a 2.5-billion-dollar industry, no longer limited to the I'm-allergic/cannot-swallow-a-pill customer. Nowadays, nearly a third of women on hormones for menopause opt for compounded. Many of the compounded pills are not covered by insurance, while big-brand estrogen and progesterone pills are.

But here's the crucial difference: due to a legal loophole, compounded hormonal products fall outside the Food and Drug Administration's purview. That means that they have not gone through the same rigorous quality control as Big Pharma products. Without FDA quality control, the pills may contain too much, too little, or contaminated hormones. These fears have been supported. In 2010, a contaminated medication supplied by a compounding pharmacy in New England triggered 750 cases of fungal meningitis, including 64 deaths. In 2013, a journalist on assignment for *More* magazine filled twelve identical hormone prescriptions at a dozen different compounding pharmacies; a laboratory analysis revealed huge variations in the amount of hormones in the pills. Manson, the Harvard doctor, said that she has seen reports of a few women taking compounded hormones who got endometrial cancer, which she suspects may have been triggered by the preparations not having enough progesterone.

Nor are compounded hormones required to have package inserts, as FDA-approved drugs must. Without warning labels, the drugs give the false impression there aren't any dangers. The lack of quality control and absence of warning labels infuriate members of the Endocrine Society, the American College of Obstetricians and Gynecologists, the American Society for Reproductive Medicine, and the North American Menopause Society.

Laws have been passed to increase oversight of compounded drugs. According to the Compounding Quality Act passed in 2013, compounding pharmacies can no longer sell a drug if the exact same item is available from pharmaceutical companies. The rationale is that it's the same drug minus the quality control. The new law also prohibits compounding pharmacies from including ingredients that the FDA does not consider safe. Estriol, one kind of estrogen, is not approved by the FDA but had been in several compounded mixtures. Also, compounding pharmacies that sell in bulk across state lines must report adverse side effects to the FDA. Doctors continue to push for package inserts that explain the potential dangers. They are fighting for change much as feminists in the 1970s pushed for warning labels on birth control pills. (Before the 1980s, birth control pills did not contain a package insert detailing the link to blood clots.)

There is some self-policing by the Pharmacy Compounding Accreditation Board, but as of October 2016, only 463 of the 7,500 compounding companies have received accreditation.

So where does that leave hot-flashing, sleep-deprived menopausal women? The 1990s ushered in an explosion of research, but we've learned much more since then. In July 2017, the North American Menopause Society issued new guidelines, updated from those that came out in 2012. The big change is this: women do not have to discontinue hormone therapy after a few years. Prior thinking was that women should stop after about five years, but recent evidence suggests that may not be necessary. Some women can stay on it for decades, reaping benefits without suffering the consequences, such as a slightly increased risk of heart disease or breast cancer. Estrogen alone, for women who have had a hysterectomy, or estrogen plus progesterone, for everyone else, is the most effective way to stave off hot flashes and to reverse painful vaginal dryness. Some women find relief by using soy or herbs or other non-hormonal vaginal lubricants, but no studies have shown that any of these are much better than placebo. The guidelines also noted alternatives, such as a new pill that combines estrogen with Baze-doxifene, a drug that taps into estrogen receptors and minimizes the risk of uterine cancer. Women on this new combination drug, called Duavee, do not have to take progesterone.

But here's the real, nagging issue: those of us old enough to be in menopause can't help but wonder if the experts are going to change their minds again.

Menopause choices highlight the ever-present uncertainty in medicine. The hope is that research, such as the studies done by Dr. Freedman and Dr. Rance, will reveal information about our estrogen-starved bodies that will lead to better therapies. But new ideas about menopause also mean that today's advice may be out-dated tomorrow. That isn't to say that the experts are fickle, though it may sometimes seem that way. They are making judgments based on the latest information, a set of data that continues to bump along.

To alleviate some of the confusion around picking the right drug—or any drug—the North American Menopause Society cre-ated an app called MenoPro. You can download it, answer a few simple questions, such as "Are your symptoms severe?" and "How old are you?" Then, with a few taps on your smartphone, you get advice and links to further information to help you figure out what is best for you.

Haseltine, with all of her degrees and knowledge about meno-pause and women's health, didn't need an app. She knew what was right for her. She considers herself a diehard feminist. Besides her academic appointment and her leadership role at the National Institutes of Health, she has also earned the American Women's Medical Association Scientist Award. But she said, "People will get into your face about their belief, but no one else's belief has anything to do with it. I looked at the information. I knew the risks. Whenever you have a C-section or a hysterectomy, it's seen by most people as emotional. So I was vilified. The whole women's movement is that if you have the information, you should be given a choice. But there is no choice. It's you should do everything naturally. Most people assumed because I was a feminist, I was for breastfeeding and natural childbirth, but I was for neither."

No doctor would suggest an elective hysterectomy. Yet Haseltine's choice, albeit unpopular, demonstrates that women today have the option to do what they want when it comes to coping with menopause—as long as they appreciate both the benefits and the risks. Haseltine studied the literature and made an informed health decision. Or, as she said, "You can use your knowledge any way that makes you feel comfortable."

12.

Testosterone Endopreneurs

THE DOGS HAD SEX in a smelly room in the basement of Grace–New Haven Hospital at Yale University. Some were castrated, some weren't, and some had been given shots of testosterone. This wasn't some animal porn movie. It was all part of a series of hormone experiments launched by Dr. Frank Beach in the summer of 1947.

When Beach arrived at Yale, he was a rising star, but he hadn't yet hit the pinnacle of his success. He had earned his PhD in psychology from the University of Chicago in 1940 and afterward served as a researcher in the department of experimental biology at the American Museum of Natural History, and founded its Department of Animal Behavior. In 1946, he moved to Yale. They gave him the title of full professor but, rather than a lavish laboratory, they offered him a small, unventilated room next to the men's room in the hospital. The space was available because no one else wanted it. Before the dogs arrived, janitors were supposed to eat lunch there, but they refused because they said the room stank like a toilet. When the canines moved in, the employees grumbled that their lavatory smelled like dogs.

The trials would span twenty years and two universities—Beach left Yale in the late 1950s for a professorship at the University of California, Berkeley. From beginning to end, Beach eyed his dogs closely, correlating sexual performance to their hormone status and grading them on their antics. One, the lowest score, was for flirtation only. Top scorers got an eight: fornication plus locking. Locking, unique to dogs and the South African fur seal, is when the penis gets stuck inside the vagina, due to an extra gland on the base of the penis that swells during sex, literally locking the couple together. After the act (which can last a few minutes to about an hour), everything shrinks and the penis slips out. Beach's article includes a photograph of a dog and a bitch having intercourse back to back, which demonstrates that a dog can do a mini-pirouette during the act without ever losing contact, thanks to his bulbus glandis.*

Among Beach's goals was to determine the effect of testosterone treatment on sexual prowess. He hoped the studies would provide much-needed information for humans. Testosterone shots were just beginning to be promoted as rejuvenation therapy for our species, stirring up controversy and mass confusion. Some scientific articles and a best-selling book claimed it was the cure for so-called male menopause, a quasi-medical diagnosis. "Yes, manhood is chemical, manhood is testosterone," wrote Paul de Kruif, PhD, in his 1945 book *The Male Hormone*. Others excoriated testosterone

---

* In his article "Locks and Beagles," which was a reprint of a talk he gave at a meeting of the Western Psychological Association in Vancouver, Canada, in 1969, he wrote, "The title of this address was not chosen lightly or impulsively. In fact it was suggested to me 20 years ago by one of my research assistants, Charles Rogers, when we were just beginning to study mating behavior in dogs. I am sure everyone knows what a beagle is and the meaning of a lock will become clear as this talk progresses."

therapy as nonsense. "Male Hormone Scant Help to Middle Aged Men" was the headline of a 1947 Associated Press article by the eminent science writer Alton Blakeslee.

Beach found that the testosterone injections worked marvelously in rats, but the results were not as predictable in dogs. Rats shot up with the hormone lusted after any caged female rodent. Male dogs sometimes rejected the bitch. That led Beach to the conclusion that the more complex the male brain, the smaller the effect of testosterone shots is on behavior. Rats were dominated by the hormones secreted by their reproductive organs; dogs not so much; humans even less so. "If the hypothesis were confirmed," Beach told scientists at a meeting of the Western Psychological Association in Vancouver, Canada, in 1969, "it would lead to the prediction that increasing neocortical complexity and dominance, such as seen in primates, would be accompanied by decreasing control of sexual behavior by gonadal hormones." In other words, maybe shots for middle-aged men—touted to jack up libido, build muscles, and energize aging brains—weren't what they were cracked up to be.

Beach probably figured that the debate over testosterone—whether the hormone repaired tired old men—would be resolved by the twenty-first century. Quite the contrary; the debate continues. In fact, the vitriol has intensified, igniting the kind of mudslinging seen among rival political candidates, not among buttoned-up medical professionals. "Hormonophobes," a doctor recently called his scientifically conservative colleagues. "Endocriminologist," said a physician of his hormone-enthusiastic peers. In the middle of this ruckus are aging men wondering whether to avoid hormone therapy or give it a shot.

The issue is not whether older men have lower levels of testosterone than they did during adolescence. They do. Beginning at age thirty, men's testosterone drops by about 1 percent a year. It's a slow leak, like a tiny hole in a bike tire that may not be noticeable until the tire is nearly flat. Along the way, as men creep into their middle years, they are straddling a bike with a limp tire, which makes the ride a lot harder than it used to be. Or perhaps one can think of the lower hormone level as more like a new steady state, like switching from a racing bike to a city cruiser.

The real question is whether these older-man hormone levels constitute a syndrome that needs fixing. And if so, does the treatment work? Is it safe?

The testosterone therapy story is remarkably similar to the story of estrogen therapy. To recap: estrogen was synthesized and sold to women as an elixir of youth, repackaged as a preventer of disease, and in a dramatic turnaround a long-term study spotlighted the pros and cons. The male version has not yet reached the finale.

In 1927, Dr. Fred Koch, a professor of physiological chemistry at the University of Chicago, along with a medical student, Lemuel Clyde McGee, eked out .0007 of an ounce of "active ingredient" from 44 pounds of bulls' testicles. They didn't know what that special component was, but a droplet injected into a castrated rooster returned the swagger to its stride and the redness to its comb. Their trial was a modernized version of Dr. Arnold Berthold's nineteenth-century backyard testicle-swapping rooster experiment, using a chemical—albeit a mystery one—instead of the whole gland. Koch and McGee confirmed their findings in castrated rats and pigs. In an article called "The Testicular Hormone," Koch, along with his colleague T. F. Gallagher, detailed the extraction process but admitted that they had held off on naming their newfound substance. "It is our feeling that until more is known about the chemical nature of the hormone no name should be given."

The following year, Adolf Butenandt, a German scientist, isolated the same stuff from men's urine, also in measly yields. He collected .0005 of an ounce from 3,960 gallons of pee. These feats were scientific advances, but far from a practical treatment. You'd need to neuter pastures of bulls or collect buckets of men's urine just to make one castrated cock crow like a fully fledged rooster.

The enigmatic hormone was ultimately christened by Ernst Laqueur, a chemist at the University of Amsterdam and a founder of the drug company Organon. He extracted pure testicle hormone in 1935 and named it testosterone—from testes (testicles) plus sterone (the chemical structure). There was griping by some fellow scientists, who argued that the moniker "testosterone" implied that it came from the testicles and only the testicles, which it doesn't. It also suggested that testosterone was exclusively a male hormone, which it isn't. The adrenal glands make testosterone, and so do ovaries—but in tinier amounts. Still, the name stuck, and with it lingering misconceptions about male hormones and female hormones.

Anne Fausto-Sterling, a Brown University anthropologist—who had written about gender and prompted Bo Laurent to start the Intersex Society—rekindled the testosterone conversation in her 2000 book *Sexing the Body*. She suggested that the term "sex hormones" be changed to "growth hormones," because that's what they do. Testosterone and estrogen affect the development not only of the ovaries, testicles, vagina, and penis, but also of the liver, muscles, and bones. Indeed, they influence nearly every cell in the body. "So to think of them as growth hormones," Fausto-Sterling once told the *New York Times*, "which they are, is to stop worrying that men have a lot of testosterone and women, estrogen."

Back in 1935, the same year testosterone was named, two scientists working independently figured out how to make the hor-

An early twentieth-century example of ideal manhood,
illustrated in Georges Rouhet and Professor Desbonnet,
L'Art de créer le Pur-Sang humain (Paris and Nancy:
Berger-Levrault, 1908). *Courtesy of The New York Academy of
Medicine Library.*

mone from scratch—the key to mass production. Butenandt, the
testosterone-from-pee researcher, was funded by the German company Schering. His competitor Leopold Ruzicka was sponsored by
Swiss company Ciba. They both accomplished in the laboratory
what the body does on its own: they tweaked a few molecules of
cholesterol and turned it into testosterone. Cholesterol (in addition

to its notorious reputation as an artery-clogger) also serves as the raw material from which the body makes a variety of hormones. The work was so groundbreaking that the two scientists shared the Nobel Prize in Chemistry in 1939.

No longer did the replenishment of manhood rely on animal glands and their questionable amounts of active ingredients. No longer did rejuvenation depend on Steinach's pointless vasectomies, the procedure that was all the rage in the 1920s. Doctors had a mass-produced medicine that would provide, as *Time* magazine proclaimed, "all the testosterone the world needs to cure homosexuals, revitalize old men."

Except it didn't. Much to the chagrin of physicians, testosterone didn't make gay men straight. Steinach—who castrated homosexuals and implanted them with a new set of testicles from heterosexual men—had also failed in that enterprise. Despite the abundant supply of testosterone and de Kruif's grandstanding book, *The Male Hormone*, sales remained tepid for the rest of the century.

To be sure, plenty of studies proved that testosterone worked wonders in men who had diseased or injured testicles, enabling doctors to help men who would otherwise never have experienced puberty or those who had suffered injuries later in life that made their energy and libido plummet. Mid-twentieth-century athletes dabbled in the hormone, too. It seemed better than amphetamines, which many competitive sportsmen were already using. Stimulants provided a frenzied, heart-racing, get-up-and-go feeling, but not the muscle mass provided by androgens (the category of hormones that includes testosterone, and promotes male sexual characteristics). The International Olympic Committee didn't launch a medical commission to fight doping until 1967; androgens weren't included until 1975.

To grow a market—one not limited to the testically impaired or to athletes—drug-makers need physicians who are willing to prescribe, patients eager to use, and a tolerable mode of delivery. Mid-twentieth-century doctors were squeamish about broaching the subject of sex with their patients. That hindered prescriptions. Despite a few articles about rejuvenation, for the most part aging men considered all the bothersome things that went along with aging an inevitable part of getting old. That hindered demand. And lastly, testosterone could only be administered by injection, which drove many potential clients away.

All that would change at the turn of the twenty-first century, thanks to a multi-million-dollar advertising campaign aimed at removing the stigma of low libido and a new easy-to-use testosterone, which came in a gel instead of a shot. Between 2000, when the gel hit the market, and 2011, the number of American men taking testosterone quadrupled, fueling a $2 billion industry. Most of the buyers were banking on the treatment to do just what the TV spots promised: revert them to their former sex-driven, thinner selves. (Another big boost to the market was the increase in direct-to-consumer drug advertisements.)

One commercial for AndroGel, the most popular testosterone gel, opens with a handsome, lean guy with a swath of brown hair pulling into a gas station in his navy convertible. A beautiful woman rides shotgun. As he gets out, he looks straight into the camera and says, "I have low testosterone. There. I've said it."

The point was well taken. If this pillar of virility wasn't embarrassed, why should anyone be?

Then we learn he didn't have low sex drive, but he was tired and moody. His doctor diagnosed low testosterone. He started using AndroGel. As he and his babe drove into the countryside, the voiceover praised the gel's ease of use and how it boosts testosterone. As the law required, the side effects were detailed, too. The voiceover sped through potential dangers, such as the possibility of cancer and heart disease, and a further danger: because the gel could rub off on cuddling lovers and children, it might give them an unwanted hormone kick. Users were advised to "discontinue gel and call your doctor if you see early signs of puberty in a child or signs in a woman which may include changes in body hair or a large increase in acne possibly due to accidental exposure." On Comedy Central's *The Colbert Report*, Stephen Colbert showed a clip from the ad and called the gel "mass-marketed, easily spread endocrine toxin."

Twenty-first-century testosterone makers also rebranded the syndrome "Low T," which is more hipster than its previous monikers: male menopause or, worse, male climacteric. Around the same time, the drug company Organon hired a doctor to conjure up a simple survey so that men could determine for themselves if they were at risk for Low T and seek treatment. Dr. John Morley, the director of endocrinology and geriatrics at St. Louis University, said he created an intentionally vague questionnaire that casts a wide net, catching men who might be low-testosterone but could also be depressed or just tired. Either way, it would expand the potential clientele. He called it the A.D.A.M. questionnaire, a first-man acronym that stands for "androgen deficiency in the aging male." Questions include, "Do you get tired after dinner?" (Did this mean conking out on the dinner table, or at a normal bedtime?) A positive response adds a point to the "needs testosterone" score. "Are you sad and/or grumpy? Have you noticed a deterioration in your ability to play sports? Have you noticed a decreased 'enjoyment of life'?"

Morley admitted recently that "it's a crappy questionnaire," which he devised in twenty minutes while sitting on the toilet scribbling ideas on toilet paper. He later cut his ties to the pharmaceutical industry and said he donated his earnings from the questionnaire, some $40,000, to his university.

Among other ploys to expand the customer base have been drug-company-sponsored articles disguised as objective news. In a tell-all essay in *JAMA Internal Medicine*, Stephen Braun, a freelance writer, revealed that he was paid by a doctor to ghostwrite glowing articles about testosterone therapy that could be pitched to consumer magazines. The doctor, in turn, had been paid by a pharmaceutical company. "The fact that the articles appeared under the byline of a physician and appeared in trade magazines with no mention of the funder behind the overall effort raised the marketing value of the pieces considerably because it is likely the readers trust information that appears to be objective and free of industry influence," wrote Braun. He quit the ghostwriting gig when the ethics of the situation started to rattle him.

All of these tactics—the bombardment of advertisements, the rebranding of male menopause as Low T, the glowing advertorials, the sloppy do-it-yourself diagnostic test—made sales skyrocket. As John Hoberman, a professor at the University of Texas at Austin, wrote in *Testosterone Dreams*, "suddenly it seemed as though the law governing prescription drugs had been suspended by a magical wish of a public seeking to fulfill a pharmacological fantasy without recourse to mundane legal mechanisms and official medical opinions."

The U.S. Food and Drug Administration does not recognize age-related low testosterone as a disease. And if an illness doesn't exist, how can there be a treatment? The FDA approves testosterone only for men who have diseases that make hormone levels plummet, such as a pituitary tumor, and defines low testosterone level as 300 nanograms per deciliter of blood or less, to be verified with two separate blood tests. And it mandates that testosterone, whether in a gel, or pellet, must contain package inserts warning that it may increase the risk of stroke and heart attack, and may lead to abuse. The Endocrine Society, the American Society of Andrology, the International Society of Andrology, and the European Association of Urology concur. The European Association of Urology issued similar guidelines.

Regardless of the FDA's advice, doctors can prescribe testosterone for anything they deem appropriate. It's called using a drug off label. The practice is not illegal, but it is not government-sanctioned. Despite the FDA guidelines that mandate testosterone-level testing twice before treatment, 90 percent of men in the U.S. who were prescribed testosterone did not have the two required blood tests, and 40 percent didn't even have one, according to a 2016 study. In an article called "How to Sell a Disease," Dartmouth researchers Dr. Lisa Schwartz and Dr. Steven Woloshin called Low T "a mass uncontrolled experiment that invites men to expose themselves to the harms of a treatment unlikely to fix problems that may be wholly unrelated to testosterone levels."

Here's what we know:

Testosterone levels fluctuate during the day, peaking around 8 a.m. and reaching their lowest point around 8 p.m. The peaks and valleys are larger for men younger than forty, but old men don't completely flatline.

Testosterone helps restore sex drive and muscle tone among men who suffer from ailments that lower testosterone, such as injury to the testicles, genetic defects, or pituitary tumors.

Testosterone increases muscle mass, as athletes have known for years.

As Dr. Alexander Pastuszak, an assistant professor of urology in the Center for Reproductive Medicine at Baylor College of Medicine in Houston, said, "any exogenous testosterone that replaces or enhances what your body produces will shut down the hypogonadal axis." In other words, taking testosterone signals the body to halt its own production, which means testicles produce less testosterone and fewer sperm. That said, testosterone is not a reliable contraceptive.

Fat men have lower testosterone levels than their fit peers. Despite claims, there are no studies proving that testosterone burns fat. A few studies found that men taking testosterone are more likely to lose belly fat, but most of these men were also dieting.

Testosterone shots and gels increase the number of blood cells. That's why some doctors who put their patients on testosterone tell them to donate blood.

Here's what we don't know:

Whether testosterone taken for years is good or bad for the heart. The data are conflicting. A *New England Journal of Medicine* study published in 2010, for instance, found that men taking testosterone were more likely to have cardiovascular problems than those not taking it. Ten men among about 100 in the testosterone group had strokes or blood clots versus one man among about 100 in the placebo group. The researchers, worried, stopped the study. A follow-up study by the same team, published in 2015 in *Journal of the American Medical Association*, found the opposite.

Most of the studies have focused on men with severely low levels of testosterone. Men with very low testosterone levels may feel better when the well is replenished, but giving testosterone to men with normal levels does not seem to have any impact. "You don't see the big improvement once men are within the normal range," said Dr. Shalender Bhasin, a professor of medicine at Harvard Medical School and director of the Research Program in Men's Health: Aging and Metabolism at Brigham and Women's Hospital in Boston. Bhasin has been studying testosterone for decades, and says that the largest differences in terms of energy and sex drive are when men go from below normal to low normal. In one of his earlier studies, he castrated rats, which quashed their pursuit of females. Then he gave them testosterone. Once they hit the normal range, their mating behavior returned, but more testosterone didn't enhance their female-seeking tendencies.

Despite claims that testosterone boosts cognition, proof is lacking. A 2017 study in the *Journal of the American Medical Association* found that a year of testosterone therapy was no better than placebo for men with low testosterone and age-related cognitive impairment.

Most importantly, we don't know what really constitutes low testosterone. Doctors say that normal levels are 300–1,000 nanograms per deciliter of blood. Dr. Joel Finkelstein, a Harvard professor of medicine, conducted a study that tried to pinpoint the dividing line between low and normal testosterone. Some 200 men between the ages of twenty and fifty took a drug that wiped out their testosterone and estrogen. Then researchers replenished the men in varying doses. Some got a placebo. The others got 1.25 grams, 2.5 grams, 5 grams, or 10 grams of testosterone daily. The trial lasted for sixteen weeks. Finkelstein found that the level of testosterone that corresponded with symptoms varied among men, so it wasn't appropriate to set one level as so-called Low T.

This problem is compounded by the fact that every testosterone-making lab has its own methods of measuring testosterone, so a man may be at 300 with one company's technique but at 400 with another. PATH, short for Partnership for the Accurate Testing of Hormones, is a group of doctors and researchers pushing for standardization of hormone tests.

The most stunning research finding contradicted what many doctors assumed. It had been thought that when a man's testosterone level declines, his estrogen level rises. This harks back to the notion in the early 1900s that estrogen and testosterone are competing hormones, which was promoted by Eugen Steinach, of vasectomy-libido fame. There's a rumor wafting about Low-T clinics today that men with low libido have extra belly fat due to an overload of estrogen. A recent study found just the opposite: men with low testosterone also had low estrogen.

And yet, there are still plenty of doctors who believe men should be able to try testosterone shots. The alternative, as they see it, is to wait decades for the results of a gold standard study that hasn't been launched yet. And, let's face it: how long can a seventy-year-old man wait for results?

Dr. Mohit Khera, an associate professor of urology at Baylor College of Medicine, reckons that since doctors are allowed to offer post-menopausal women estrogen for hot flashes without testing their hormone levels, men should be offered a trial of testosterone in the same way. "For some reason, we think that with men, we should not pay attention to symptoms and pay more attention to the numbers. It doesn't make a lot of sense," said Khera, who is also a consultant for two testosterone makers, AbbVie and Lipocine.

The difference is that estrogen has been proven to prevent hot flashes. Testosterone has not been proven to boost libido and fight fat among men who are not severely deficient in the hormone. "You can't really recommend treatment when you could be doing harm," said Harvard's Finkelstein.

What infuriates Finkelstein and his ilk are the anti-aging doctors who tout the benefits of hormones to boost longevity and quality of life. Endocrinologists find the crass promotional tactics reminiscent of the organotherapists of years past: the unscrupulous doctors and charlatans of the 1920s who peddled goat and monkey glands.

At a meeting on hormones sponsored by the American Academy of Anti-Aging Medicine in September 2016, Dr. Ron Rothenberg touted the benefits of testosterone for aging men. Rothenberg, seventy-one, is the medical director of the California HealthSpan Institute in Encinitas, California. He takes hormones for rejuvenation and so do his clients. Photos of him surfing decorate his website. Rothenberg is short and energetic, with chiseled arms and an orange complexion. He pranced back and forth across the stage like an evangelist as he preached to a packed audience of physicians in a ballroom at the Hyatt Regency in Dallas.

"How do you define what's a deficiency?" he asked, without waiting for a response. "An older concept was that if you are normal for your age, you are normal. If you are eighty years old and getting glasses to correct your vision—for a normal eighty-year-old? It's silly, really. Testosterone is getting lower every year worldwide. It'll be zero eventually. It's like a disaster movie."

Rothenberg blamed the medical establishment and the media for mocking Low-T. There has been a slew of articles in the past few years slamming the Low-T industry. While other doctors worry about the potential dangers of taking testosterone, Rothenberg worried about the health hazards of living with low testosterone. He argued that low testosterone increases the risk of heart disease, which is the opposite of what other doctors say. He also claimed that men with low testosterone have an increased chance of Alzheimer's disease. (There are no good data to back his assertions.)

Rothenberg, like other doctors at the meeting, declared that they treat the whole patient, not a list of laboratory results. "I don't have my nose stuck in the lab. Let's say it [testosterone] starts at 300 and comes back at 500, how do you feel? Great, good, let's say it starts at 300 and it comes back at 1100, that's fine too. I'm not trying to hit an exact number."

After Rothenberg's morning session, a throng of doctors gathered around him at the podium, as if he were J. K. Rowling at a Harry Potter book signing. I joined the crowd because I wanted to ask Rothenberg why there were so few endocrinologists at this hormone conference, and what seemed to be an abundance of former emergency room doctors.

"ER docs," he responded, "are more about diving in without knowing everything, not to compulsively need to know everything." They are, he said, willing to try.

There was something else strange about this conference compared to other medical meetings I'd been to, but I couldn't figure it out right away. Then, during a break between sessions, I ran into Dr. Roby Mitchell, who has an MD and a PhD. He nailed it: the meeting was more like an infomercial than an educational seminar. Most medical meetings allow time for debate, with doctors enjoying the intellectual banter. The information here was presented as dogma without time or inclination for challenges. Mitchell admitted that it's marketing, but "as a consumer, it's your job to filter out what's useful and what's bullshit."

The American Academy of Anti-Aging Medicine is not recognized by the American Medical Association, nor are its boards certified by the American Board of Medical Specialties. To become board-certified in endocrinology the traditional way, a doctor must complete two or three years of intensive training after residency and then sit for an exam. To become board-certified in endocrinology by the American Academy of Anti-Aging Medicine, a physician must complete four modules, each of which involves eight hours of online learning, and 100 hours of continuing medical education credits (provided by attending the Academy), submit three case studies, and pass a written exam. The certification is in Metabolic and Nutritional Medicine, but includes information about hormones, a spokesperson for the Academy said.

At the Dallas meeting, I saw a doctor going into the exam room and asked him why he would spend time and money on a test that wasn't recognized as board certification by the medical establishment. He looked at me as if I had twelve heads, as if I were the one at the party who didn't get the joke.

"At least you get a paper to hang on your wall," he said, with a laugh. "Patients like that."

In *Selling the Fountain of Youth: How the Anti-Aging Industry Made a Disease Out of Getting Old—And Made Billions*, Arlene Weintraub wrote that "these capitalists have constructed a giant new industry by taking advantage of an entire generation's deep-seated aversion to getting old." She's right. I would add that that it's not just this generation, but a longing for rejuvenation that has spanned the history of hormone therapy.

Many urologists and endocrinologists believe that, while testosterone may be abused among the worried well, it is underused among men who truly need it. There's no evidence. Without mass screening, no one knows if men with severely low levels are being overlooked. In 2015, the American Academy of Family Physicians published two opposing opinion pieces to this prompt: Should family physicians screen for testosterone deficiency? Dr. Adriane Fugh-Berman, a Georgetown University professor who writes a blog called "Pharmed Out," noted that "testosterone testing leads to testosterone treatment, which is inappropriate for the vast majority of patients." In the other camp is the University of Michigan's Dr. Joel Heidelbaugh, who wrote that, while physicians should be cautious whom they treat, "it is evident that many men likely have untreated symptomatic testosterone deficiency." Screening should be offered, coupled with laboratory tests and a discussion of the potential risks and benefits.

In the meantime, more than 5,000 men who claim that testosterone therapy caused their heart attacks, strokes, or blood clots are suing the industry. There are no statistics on the number of men who have died or gotten sick from taking testosterone, because it's hard to prove that it's the therapy that triggered the problem; it might have happened anyway. The lawsuits have been gathered together into one multidistrict litigation in Chicago. One judge heard testimony from eight cases deemed representative of the diversity of arguments. Thus, the drug companies do not have to rehash their defense for each one of the thousands of cases. This is not a class action suit, but the results will shape the way the rest are individually tried. The first cases were heard in the summer of 2017 in Chicago. On July 24, a federal jury decided that AbbVie should pay $150 million in punitive damages to an Oregon man who had had a heart attack and accused the company of misrepresenting the risks, though they found in favor of AbbVie against allegations that the company was negligent and that it had not provided adequate warnings.

Frank Beach, the dog-sex researcher, may never have imagined that the kernels of truth he uncovered would foster a multi-billion-dollar industry. He died in 1988. Long before that, he expanded his scope beyond testosterone, to the thyroid and adrenals and other hormones that impact behavior. He became famous as a pioneer in the field of behavioral endocrinology. For him, the impetus was a scientific quest to unravel the mysteries of the endocrine system.

Beach was a bear of a man, with a grizzly white beard, a slight paunch, and scruffy attire. He was funny and gregarious and never lost his Kansas down-to-earthness. He taught high school English before and during his doctoral work. One day in the late 1950s, when he was in his office at Yale, a graduate student named Peter Klopfer knocked on his door. Klopfer had been advised to introduce himself to the distinguished researcher, and expected someone Ivy-Leaguish, wearing a tweed blazer and khakis and sitting behind a grand mahogany desk. Not Beach. He was leaning back, legs up on his desk, wearing a ripped, stained T-shirt and pounding down a Pabst beer. "I was just shocked, shocked to my bones," Klopfer recalled years later. The office was decorated with photographs of erect animal penises. "There's this famous guy who looked like a hobo from the back streets of New York City."

Beach suggested they retire to the local pub to chat over beers and pizza. And so they did. "Beach was one of the most brilliant people I've known," said Klopfer, now a Duke University professor emeritus. "There was this incredible difference between his appearance and his intelligence. It took me years to appreciate."

Klopfer was in the biology department, but often regretted that he hadn't switched to psychology just to be under Beach's tutelage. Klopfer would, however, follow in Beach's footsteps, immersing himself in animal studies. He studied maternal–fetal bonding, clues that led to research into the hormone oxytocin. And while their fields were vastly different, Beach and Klopfer were each at the forefront of specialties that were grounded in solid research and then exploited in all sorts of endopreneurial ways.

As for Beach's dogs, one mutt named John Broadly Watson topped the charts with a 100 percent approval rating, meaning that none of the bitches rejected him. He was—a surprise to Beach—the least dominant of the five males.

13.

Oxytocin: That Lovin' Feeling

D R. PRUDENCE HALL gave oxytocin, a hormone, to her son before he hung out with his friends at a college bar. All of the girls flirted with him, supposedly seduced by the powers of the endocrine aura. Another time, her daughter took oxytocin before a graduate school exam and said she was more relaxed and focused than she would have been without the drug. Hall, who is the medical director of the Hall Center, a health clinic on Wilshire Boulevard in Santa Monica, California, sells oxytocin to her patients who get social jitters before parties, who have lost their sex drive, or who just don't feel like their old friendly, loving, trusting selves anymore.

She shared her oxytocin candy with her publicist, her assistant, and me before we sat down to talk. It looked like opaque white pebbles but tasted like sugar cubes. We slipped them under our tongues for the optimal rush. That method, she explained, drives the drug to the brain quicker than the oxytocin nasal sprays you can buy online.

Dr. Hall is trained as an obstetrician-gynecologist but, in her new broader role, she takes care of men, too. She has flowing blond hair and a soothing way of speaking. On the day we met, she wore a purple tunic with a long necklace of crystals and a tassel. The clinic is decorated with Thai teak furniture, cozy couches, and nature images on saffron-colored walls. Hall looked like someone who would run a meditation retreat. The whole place felt like a spa rather than a doctor's office.

There's a shop in the center of the clinic that sells, among other herbal remedies, Hall's own line called Body Software. The pink bottle is Secret of Feminine Radiance; the green one is Prostate Protect. There's also Mega Adrenal and Super-Adrenal, which are said to boost motivation. Hall has been on television, on both *Dr. Phil* and *Oprah*, where she was interviewed by Dr. Mehmet Oz before he had his own syndicated medical program. She counts Suzanne Somers, the actress and diet book author, and Sarah Ferguson, the Duchess of York and former Weight Watchers ambassador, among her clients.

"Do you feel it? I'm feeling a little more intense," said Dr. Hall, as we waited for the oxytocin to kick in. Then she leaned closer to me and added, "I want to look into your eyes."

Her publicist said she felt it too. And she leaned toward me. I didn't feel anything.

Oxytocin—not to be confused with oxycodone, the narcotic— is a brain hormone. During labor, oxytocin prompts the womb to contract, pushing the baby through the birth canal. Afterward, it triggers the breast ducts to squeeze out milk. The synthetic form of oxytocin, Pitocin, kick-starts labor, providing an extra oomph to get the womb pumping. But recent research has rebranded this heady substance into something more marketable than its prior matronly connotations. Oxytocin has been said to foster the bonds between mothers and newborns and between lovers, to bring on erections and orgasm and ejaculation, and enhance mind-reading. It's unclear whether or not that all happens at the same time, or in what particular order. Oxytocin has also been linked to trust and empathy. One small study showed it boosted compassion between Israelis and Palestinians. But here's the rub: among the flood of studies (upward of 3,500 oxytocin behavior studies in the past decade), oxytocin has been tied to trust but also to distrust; to love but also to envy; to empathy but also to racism. That ought to flummox a potential oxytocin customer.

The first clues to oxytocin's powers emerged in a 1906 study by Henry Dale. Dale had just graduated from university and, while he was preparing his applications to medical school, he was appointed director of London's Wellcome Physiological Research Laboratories. The lofty title—director of a lab at such a young age—came with a caveat; he was tasked with investigating the science behind ergot, a fungus. For Dale, the ergot assignment was an insult. "I was, frankly, not at all attracted by the prospect of making my first excursion into the ergot morass," he wrote. Ergot was a folk medicine used by midwives to speed childbirth and cure headaches. Other physiologists were investigating the internal secretions of the pituitary and the thyroid and the pancreas—serious subjects, with the potential for making a mark on the field.

Dale did the obvious experiments, injecting ergot into a menagerie of animals—cats, dogs, monkeys, birds, rabbits, and rodents. He recorded the increased blood pressure and muscle contractions it provoked. Then he added a twist, giving some animals a combination of ergot and adrenaline, the fight-or-flight hormone. Ergot

stalled the adrenaline charge. These findings led to the first generation of blood pressure medications.*

In the middle of his so-called ergot morass, between shooting
up rodents and monkeys with the folk remedy, Dale gave a shot of
dried ox pituitary to a pregnant cat. Perhaps he was inspired by
Harvey Cushing; the pioneering neurosurgeon and endocrinologist
was at that time on a lecture tour, talking about the pituitary and
its life-altering juices. Scientists were beginning to realize that the
two lobes of the pituitary contained completely different chemicals.
Dale used the posterior lobe, and, lo and behold, the cat's uterus
contracted. In his forty-three-page article "On Some Physiological
Aspects of Ergot," Dale doesn't say what drove him to acquire a
pituitary from an ox, nor why he gave it to a pregnant cat, nor why
he used the posterior but not the anterior lobe.

Pituitaries and their secret secretions were the talk of physiologists then. Dale's lengthy article included one graph—one chart
among twenty-eight—which illustrated the increase in uterine
pressure after the shot of dried pituitary. His conclusion summa-
rizes the functions of ergot, but he slips in this line: "The pressor principle of the pituitary (infundibular portion) acts on some
constituent of plain muscle fibre other than that which is excited
by adrenaline." To put it simply: some substance produced in the
posterior lobe of the pituitary squeezes muscles.†

---

* Dale would go on to win the Nobel Prize in chemistry in 1936 for his work on chemical trans-
missions in nerve impulses. His son-in-law, Lord Todd, won the Nobel in chemistry in 1957.

† Dale's article, published in the *Journal of Physiology*, went into great detail about the care
he devoted to his experimental animals. He was the assistant who had killed the mutt at the
center of the so-called Brown Dog Affair in 1903, so he was defensive when describing his
current study.

Dale's discovery remained buried within the journal, over-looked by the medical community. It was, in many ways, reminiscent of Dr. Arnold Berthold's 1848 rooster-testicle study. The import of both experiments was ignored for decades, until curious doctors excavated the past to create a path for the future. Starling and Bayliss were the ones who rediscovered Berthold's work and popularized the concept of hormones. Dale's work languished until the 1940s, when a team of doctors picked up where he had left off and confirmed that injected posterior-lobe pituitary extract makes the womb of a pregnant animal contract. Then they discovered the link to breast milk, as recorded in a letter to the editor of the *British Medical Journal* in 1948 describing how beads of milk trickled from a laboring woman's nipple with each contraction (she was still nursing her previous child at the time she gave birth to her next). Could the same chemical that was squeezing the uterus be provoking the milk to flow? As it turned out, it was. The mysterious pituitary hormone was finally isolated and synthesized in 1953, earning its discoverer, the American scientist Vincent du Vigneaud, the 1955 Nobel Prize in Chemistry. It was christened oxytocin, from the Greek for "swift birth."

The isolation of the hormone spawned a trove of studies into its nature. It is made in the hypothalamus, an almond-sized gland deep within the brain; from there, it slips down to the posterior lobe of the pituitary, which releases the hormone in bursts.

Around the same time, another group of scientists was investigating the chemical basis of maternal bonding, a topic that seemed very different from the oxytocin studies. But soon enough, the two fields would, if not quite merge, then overlap.

Scientists studying mother-child love wondered what, if anything, compelled a new mother to nurture and protect her new-born. Was it the smell of the baby? The sound of its first cry? The sight of a mini-me? Or a hormone?

Emerging animal research suggested there was a window in which maternal love developed. A goat study by Peter Klopfer—the Yale University student who had been so shocked and impressed by Frank Beach—showed that if you whisked a newborn away imme-diately after birth and returned it just five minutes later, the mother would reject it, treating the kid like a stranger, head-butting it and pushing it away from her nipples. The same was true for rats, who would reject their newborn if separated from it for a few minutes after birth. This suggested that if there was a hormone controlling the maternal bond, it must spike during childbirth and plummet quickly thereafter. Klopfer had read a few articles about oxytocin. He knew that it surges during pregnancy, contracting the womb and the milk ducts, and disintegrates rapidly—meaning that the level of the hormone in the blood rises and falls dramatically. Could this very same substance, oxytocin, be responsible for cementing the bond between mother and child?

Klopfer began his mother-baby goat studies as a graduate stu-dent in the 1950s, working on a farm outside New Haven. He got tired of sleeping in the barn in order to snatch the newborns right after birth, so he brought a few pregnant goats to a house he sublet from a professor on sabbatical. He turned the living room into a makeshift barn by lining the floor with turf. It worked well until the professor returned home without warning, and was aghast at the new décor and the resident gaggle of pregnant goats.

Klopfer soon left Yale for a professorship at Duke University. He bought a house in North Carolina with a backyard big enough for all of his animals, which enabled him to expand his mother-newborn studies. In an unexpectedly fortunate turn of events, Klopfer hired Cort Pedersen, a recent Duke graduate, to paint his house. Pedersen was doing odd jobs to make money while applying to medical school. The two chatted about goats and mother-child bonds, and Klopfer shared his idea that oxytocin might have something to do with it. Pedersen asked if he could join Klopfer's lab. So began a decades-long friendship and scientific collaboration.

One of the things that happens during childbirth, and only during childbirth, is that the cervix and vagina stretch tremendously. This physical expansion had been shown to trigger oxytocin release. Pedersen devised a balloon-like contraption that expanded the vagina. The goal was to provoke a surge of oxytocin in goats that weren't mothers and observe whether they bonded with random newborns. Typically, a virgin female rejects a strange kid.

It worked. Two females fitted with the balloon device nuzzled newborn kids and even allowed the little ones to suckle at their milk-free breasts. The other goats were hostile to the strange kids. The study was completed before Pedersen even started medical school, though the results were never published. A few years later, in 1983, their findings were confirmed by a University of Cambridge team. In that study, published in the prestigious journal *Science*, eight of ten vagina-stimulated female sheep nuzzled and licked random newborns. Eight of ten sheep without the vagina gadget head-butted the strange lamb. To stimulate the hormone conditions of pregnancy, the researchers also primed non-pregnant ewes with estrogen and progesterone; they found that these hormones also enhanced bonds with newborns, but not to the extent that oxytocin did. The estrogen and progesterone took a few hours to kick in, rather than a few minutes, and only worked on half the sheep. "The mechanism by which vaginal stimulation permits the immediate expression of maternal behavior in sheep is not known," the investigators concluded in their article. "However a discussion of the importance of vaginal stimulation for maternal care in goats suggested a role for oxytocin, the release of which may be of some importance, since its injection directly into the cerebral ventricles stimulates maternal behavior of non-pregnant rats." In other words, evidence was growing that oxytocin helps cement mother-child bonds. Pedersen would continue with the oxytocin research and become a professor of psychiatry and neurobiology at the University of North Carolina—and an oxytocin expert.

After Pedersen finished medical school, he conducted what Klopfer described as "this brilliant study." Klopfer had injected oxytocin into the bodies of virgin female rats and male rats, to see whether it would prompt maternal behavior. It didn't. He had a hunch that the hormone was broken down before it crossed into the brain. "Cort was the one who nailed that," said Klopfer. Pedersen injected a tiny amount of oxytocin directly into the brain of a virgin female rat, in an area called the lateral ventricle, near the hypothalamus, where oxytocin is made. Normally virgins are hostile to newborns, but those given oxytocin in the brain licked and nuzzled the newborns. They even exposed their nipples, as if they were trying to breastfeed. Further studies by other scientists showed that interfering with the oxytocin pathway in pregnant rats blocked the postpartum onset of maternal behavior. The new mothers didn't nurture their pups. Some were downright nasty, nudging the little ones away.

A mother goat without the surge of oxytocin headbutting her kid. *Courtesy of Peter Klopfer.*

Further experiments investigated whether oxytocin played a role in other behaviors tied to love or to nurturing. A few squirts of oxytocin into the brains of female rats encouraged them into a stance known to be receptive for sex: butt high. The females who weren't given oxytocin remained aloof. Male rats given oxytocin spent more time sniffing and self-grooming, but as oxytocin injections didn't speed ejaculation, the researchers concluded that oxytocin might enhance social interactions but not actual sexual performance. It was also found that the hormone hits odor receptors, possibly boosting a mother's sensitivity to her newborn's aroma.

These studies, in turn, prompted another group of investigators to see if oxytocin might account for the differing behavior of three kinds of voles (little brown furry rodents). After prairie voles have sex for the first time, they remain together for life. They make babies; they groom each other; they share childcare fifty-fifty. But their cousins, meadow voles and montane (or mountain) voles, go from relationship to relationship, never settling down. Sue Carter, director of the Kinsey Institute, found an uptick in oxytocin level in prairie voles after they have sex but not in the other kinds of voles, which suggests that oxytocin is what makes the difference between "till death do us part" and "I'm outta here." But there was a twist to the seemingly loyal, oxytocin-rich prairie voles: they stuck around to raise the kids but cheated on their spouses. Carter ran DNA studies and found that the males fathered lots of children both within and outside the partner bond.

Studies elsewhere investigated the role of oxytocin in contracting other muscles, such as those of blood vessels in other parts of the body. In 1987, Stanford University scientists recruited a dozen women and eight men who agreed to masturbate while blood samples were taken, and found that oxytocin levels spiked right at the onset of orgasm. It's hard to draw conclusions as to whether oxytocin brings on orgasms or vice versa.

How does this translate to humans, and the way we think and feel? In 1990, Carter compared twenty breastfeeding mothers with twenty who weren't breastfeeding. The breastfeeders had higher levels of oxytocin, as expected; they also were calmer, an unexpected finding. Carter suspected that oxytocin prompted feelings of peacefulness, which help nursing mothers endure the monotony of feeding. Other studies suggest that oxytocin has something to do with making people feel good generally, not just after orgasms or while breastfeeding.

The really dramatic experiment—the one that exploded the topic away out of the science journals and into the headlines—was one that used oxytocin in a game of trust. Here's how it was played. Volunteers were paired. Each player received 12 units of play money. One player, labeled the "investor," could choose to keep his money or give 4, 8, or 12 units to his partner, labeled the "trustee." Whatever the trustee got would be tripled. So if the trustee was given 12, he'd end up with 48 (the 12-unit transfer tripled to 36, plus his original 12). The trustee then had the choice to hand back any amount, or none, to the investor. There were four possible outcomes: both players could end up with more money than either had before the game; only the investor would end up with more; only the trustee would end up with more; or both would finish where they started. The researchers, a combined team of Swiss and Americans, figured that if the investor didn't trust the trustee, he would keep all the money, but if he did trust him, he'd hand over all 12 units and assume he'd get back at least the same amount and maybe even a little more. They found that when the volunteers inhaled oxytocin, they were more likely to fork over the money. The findings were published in 2005 in the scientific journal *Nature*.

The discovery that sniffing oxytocin enhanced trust made headlines in the U.S. and Europe. It also gave rise to the moniker "the moral molecule," along with a slew of self-help books (35 *Tips for a Happy Brain*, for instance), a clothing spray called Liquid Trust, and a TED talk promoting oxytocin's trust-building effects that has more than 1.5 million views to date. Paul Zak, PhD, one of the researchers and the TED speaker, proclaimed that countries with a higher proportion of trustworthy people are more prosperous, so understanding the biology of trust will alleviate poverty. Zak, who is a professor at Claremont University and the author of *The Moral Molecule*, is handsome and entertaining, with a shock of blond hair and chiseled features. "Is it really a moral molecule?" he asked during his

TED talk. And then he quickly answered: "We've run studies that it increases generosity, increases donations to charity by 50 percent." Then he walked into the audience offering spritzes of oxytocin.

Zak once blogged that oxytocin "makes us care about our romantic partners, our kids, and our pets. But here's the weird part: when the brain releases oxytocin we connect to complete strangers and care about them in tangible ways. Like giving them money." He wrote another piece about being stalked by a former girlfriend. Why? Their oxytocin was out of sync, which made his love fade and hers persist.

In reality, oxytocin isn't the sensation it's made out to be. In the original trust study, only six of the twenty-nine oxytocin sniffers gave all of their money units away. Unlike Zak, who's made a career out of extrapolating the original hormone findings, his co-authors considered that study intriguing, but not conclusive. The results have not been replicated but that could be because subsequent trials were flawed, University of Zurich's Ernst Fehr, one of the original Trust Game researchers, told the *Atlantic*. "What we're left with is a lack of evidence," he said. "I agree that we have no replications of our original study, and until then, we have to be cautious about the claim that oxytocin causes trust." Still, the facts haven't gotten in the way of a good story. Further studies linking oxytocin nasal spray to love and trust have garnered media attention. Other studies found the opposite, that oxytocin decreased trust and increased racism. The contrary findings are explained away by the supposition that oxytocin doesn't simply boost good feelings but rather amplifies whatever you're feeling in the moment.

"It's a great story that oxytocin, the same hormone that increases monogamy in voles and is involved in lactation and facilitating birth, is making you send money to strangers," said Gideon

Nave, an assistant professor of marketing at the University of Pennsylvania's Wharton School. "Somehow this has become a coherent story. If you take a lot of dots, you can always draw a line and tell a good story, even if it's just your imagination. It's well written and heavily popularized by the press."

Nave studied the studies. He's not a hormone expert; he's a statistician. He found that, for the most part, the studies were too small, too biased, or too sloppy to prove anything. Most could not be replicated, meaning that the results might have been simply a chance finding, a fluke. What's more, Nave trawled through the desk drawers of oxytocin researchers and found some human studies showing that oxytocin didn't influence behavior. Those studies were never published. Professional journals (and the journalists who rewrite scientific articles for newspapers) tend to prefer positive findings. But it's precisely the so-called negative studies that provide a nuanced account of reality.

Leaving aside study bias, it may seem that there's not a whiff of evidence that oxytocin impacts human behavior, but that doesn't mean the hormone doesn't do anything. It just means we don't have proof yet. Skeptical endocrinologists assert that the studies have been extrapolated a little too much. The debate is reminiscent of a letter that Harvey Cushing received nearly a hundred years ago after one of his lectures on the pituitary. "It is pathetic if not disgusting to witness this endocrine orgy now rampant in our profession, much of it the result of abysmal chaotic nonsensical ignorance, much of it alas the result of commercial greed," wrote Dr. Hans Lisser, head of the University of California San Francisco's ductless gland clinic. "Endocrinology is fast becoming a mockery and a disreputable business and it is high time that some honest fearless words were uttered." Larry Young, director of the Silvio O. Conte

Center for Oxytocin and Social Cognition at Emory University, said it's not so different these days. There's a lot of good stuff mixed in with the bad. After reading Lisser's letter to Cushing, he said it's "an oxytocin orgy out there."

Young is part of a cadre of neuroscientists, including New York University's Robert Froemke, who are conducting careful studies of oxytocin, trying to understand the hormone by zeroing in on the precise location of oxytocin receptors in the brain. Froemke's work built on Young's work, and also on a 1983 study of ten nursing women that found that merely the sound of an infant crying bumped up oxytocin levels. "From a neuroscientific perspective, these infant cries come into the ear and are processed in the brain by the auditory system," he added. He found an abundance of receptors in the left auditory center, more so than on the right side, and reported that mice with blocked oxytocin receptors in the left auditory center did not respond to crying babies the way that mice without the blockers did. Like many of his colleagues, he does not believe that oxytocin sparks the maternal bond, but rather that it enhances incoming information. As he put it, "These things take on a new richness." He explained it this way: "Everyone has been on an airplane with a crying baby and we have different experiences. Some find it annoying, obviously. But often there are some women who start lactating at the sound of the baby crying. That, biologically, is really amazing." In other words, oxytocin may act by augmenting lurking feelings.

While Froemke has focused on hearing, other scientists have studied social responses, hoping that, as they decode the effects of oxytocin on the body, they may be able to manufacture therapies with real impact. Because some of those effects suggest that oxytocin enhances social skills, it has been tested as a treatment for autism and schizophrenia. Findings so far have been mixed. A key issue is that injecting oxytocin into the brain—which works in rodents—is not something that can be done in humans, even experimentally, and no studies have yet proven that inhaling oxytocin nasal spray boosts brain levels of the hormone. "It might be helpful, but it's too early and people are too optimistic at this point about the intranasal route," said Young, the Emory oxytocin researcher. "Maybe it's harmless," said Young. "But I personally don't think we are at the point where we can be really confident. Even if some of the papers were true, the effects are relatively small. I can't imagine you take a sniff of oxytocin and go to school and come home and take another sniff and it will improve functioning."

Young added that, even if these early studies have flaws, we shouldn't give up on the whole thing. He believes that as we decipher the nitty-gritty of how oxytocin works, we may find novel treatments to help people with autism or social anxieties. "It's not FDA approved for anything yet," he cautions. "Physicians can get it and parents are begging for it for their children."

At issue isn't whether oxytocin plays a role in birth, sex, and behavior. It does. What we—potential customers, scientists, and journalists—need is clarity. Within the vast ocean of studies are pearls of evidence, tantalizing clues that will guide future researchers to figure out what oxytocin really does and how, if at all, we can tap its potential. "Some of the things they say may be true: that oxytocin is involved in love and sex and decreasing anxiety and stress and all that kind of stuff," said Pedersen, the University of North Carolina researcher. "It's just that translating it into a useful treatment is going to require an awful lot more work."

Dr. Prudence Hall isn't concerned by the negative talk about oxytocin. As she said, she's not a researcher, she's a clinician, and she knows what works for her patients. She doesn't care what the data says about how much of the hormone gets to the brain. She's seen the impact of her sublingual tablets. As we finished talking about oxytocin, Dr. Hall hugged me. Then her publicist hugged me. Then her assistant hugged me. As I was about to walk away, Hall added, "Hugging helps oxytocin too."

14.

Transitioning

MEL WYMORE STARTED TAKING TESTOSTERONE just before meno-pause hit. As it turned out, he and his son went through puberty together. His son developed an Adam's apple and a deeper voice first. "I trailed him," Mel said.

Mel had been divorced for nearly ten years when he made the decision to start transitioning his appearance. "I sat down with the kids down and I pulled out an album of my childhood. I said, 'You guys know that I'm not the typical mom because I date women and you've seen me cut my hair short and I'm discovering there is a boy inside of me that I've been hiding. I'm going to let that boy out.'"

Mel switched to a masculine wardrobe, restyled his hair into a traditional man's cut, and wrapped his breasts to flatten them. "One of the first things I did was bind my breasts. It was such a relief to get rid of the bras and to masculate my feminine qualities."

His children were supportive; they were twelve and fifteen

at the time. But Mel said they had no idea what was to come. Nor did he.

Mel, like others in the transgender community, believed with a deep-seated conviction that his female anatomy did not conform to the way he felt inside. That is not the same as sexual orientation, which is about desire. People in the trans community like to say that sexual orientation is whom you want to go to bed with; gender identity is who you go to bed as.

According to global surveys, between 0.3 and 0.6 percent of people worldwide consider themselves transgender. A 2016 U.S. questionnaire yielded similar results, which would give a figure of at least 1.4 million transgender American adults. These numbers do not take into account people who are afraid to admit how they feel. It's no surprise that the rate of people who identify as transgender appears to be higher in places with anti-discrimination laws.

The statistics, along with a spate of media (articles, books, documentaries, and television shows with transgender characters) may give the impression that being transgender is a twenty-first-century creation. But accounts of men and women feeling they were born into the wrong body have existed for centuries. In generations past, that meant changing wardrobe and adopting a new name. The rise of plastic surgery in the early twentieth century allowed a few people to have operations to remove or alter unwanted organs. In 1930, for instance, the Danish painter Lili Elbe began the first of four operations that comprised castration, molding a penis into a vagina, and implanting ovaries and a uterus.* The big difference between then and now is that hormone therapy has provided a safe means of

---

* Elbe died September 1931, during the operations that included making a vagina and implanting a uterus.

Christine Jorgensen, November 4, 1953. The original caption read: "Actress Christine Jorgensen in her first picture in a bathing suit." *Bettmann/Getty Images.*

making the physical transition complete. And it all began with the availability of testosterone in 1935 and a synthetic form of estrogen in 1938.

On December 1, 1952, New York's *Daily News* broke the story of Christine Jorgensen, formerly George Jorgensen, an introverted twenty-six-year-old soldier from New York City who transitioned with the help of surgery and hormones. "Ex-GI Becomes Blonde Beauty," ran the banner headline. The bottom half of the page showed two photographs side by side: a profile of Christine with a short Marilyn Monroe–like bob next to a headshot with a buzz cut partially covered by an army garrison cap. Christine Jorgensen was the Caitlyn Jenner of the 1950s, which is to say that she was not the only person who underwent transgender treatment, but she was the most vocal.

In the years prior to the operation, Jorgensen considered psychoanalysis. Jorgensen thought about taking testosterone, an idea sparked by reading Paul de Kruif's popular book *The Male Hormone*. But Jorgensen didn't believe either would change the entrenched sense of a feminine self. The book, which touted testosterone for men, got Jorgensen thinking just the opposite. "Could the transition to womanhood be accomplished through the magic of chemistry?"

Jorgensen duped a pharmacist into supplying estrogen pills, even though a prescription was required, by claiming to be a medical assistant in training (which was true) who needed 100 estrogen pills for an animal experiment (which wasn't true). The label said, "Not to be taken without the advice of a doctor." Jorgensen took one tablet every night. After the first week, Jorgensen's breasts felt tender. Jorgensen also felt more rested than ever—perhaps due to increased happiness rather than because of the drugs, Jorgensen wrote years later in an autobiography.

Jorgensen found a sympathetic doctor in New Jersey willing to refill the prescription and monitor the effects. After about a year on the hormones, he mentioned a surgeon in Sweden who did transgender surgical operations. So Jorgensen set sail for Europe, intending to spend time with relatives in Denmark before heading to Sweden. Ultimately, Jorgensen found a Danish doctor, Dr. Christian Hamburger, willing to perform the surgery at no cost.

(Because it was considered experimental, it was covered by the government, and Jorgensen's Danish parents qualified her to receive it.)

On September 24, 1951, Jorgensen had the first of three operations to remove her testicles and mold her penis into a vagina. Three months later, after the final operation, the news hit the American press.

A front-page story in the *Chicago Daily Tribune* said Jorgensen's parents received a letter from their son "telling them how surgery and injections changed him into a normal woman." Another article in the *Austin Statesman* included a telephone interview with Jorgensen. The reporter asked her if her hobbies were male or female. "I mean are you interested in say, needlework rather than a ball game?" Jorgensen responded, "If it's a normal female interest then it interests me."

Jorgensen returned home a media sensation and began a career as a nightclub entertainer. By her own admission she couldn't sing or dance, but, as she described it, "I somehow skyrocketed from a devastating failure in Los Angeles to star status in a world-famous nightclub and my name in lights on Broadway, all in less than a year."

Many Americans read the articles and watched Jorgensen perform with prurient fascination. Dr. Hamburger's office in Denmark was flooded with letters from desperate foreigners requesting surgery. He referred the Americans to Dr. Harry Benjamin, an endocrinologist who specialized in issues of gender and sexuality at his offices in New York and San Francisco.

Benjamin would go on to write *The Transsexual Phenomenon*, a landmark book that popularized the notion that transgender identity has a biological cause and is not due to psychological trauma or bad parenting, as was previously thought. He also wrote the introduction to Jorgensen's autobiography, describing Jorgensen's healthy, normal family with a charming mother and a father who served as the proper role model for a son. Benjamin explained that scientific evidence suggests the feeling of being male or female is created in the brain of the developing fetus. While the details were still fuzzy, he said, "it is a possibility strong enough to bring the psychoanalyst's dictum of childhood conditioning, as the only cause of transsexualism, close to dethronement." He also distinguished between "transsexuals," who want to switch gender, and "transvestites," who want to dress as the other gender but do not have the conviction that they are in the wrong body. (The term "transsexual" was used until it was replaced by transgender in the 1990s.) Benjamin referred many of his patients seeking surgery to the team at Johns Hopkins that had operated on intersex children. Dr. Howard Jones, the gynecological surgeon there, was eager. Years later, he said that as soon as he read the newspaper reports about Jorgensen, he figured he could do transgender surgery if the Europeans were doing it.

Jones saw the surgery as a challenge, perfecting techniques and creating operations that were not yet in textbooks. He believed that Johns Hopkins had the ultimate team of experts to deal with the transgender population: endocrinologists, psychologists, urologists, and plastic surgeons who were accustomed to working together because of their experience with babies born with ambiguous genitalia. The Johns Hopkins Gender Identity Clinic officially opened in 1966, the first transgender clinic within an established medical setting. (They had also treated a few transgender patients in the 1950s.) The Hopkins team required that patients cross-dress for two years, and pass a psychological evaluation before starting hormone therapy and surgery—a protocol that was based on assumptions rather than on scientific evidence.

Jones recalled that his wife, Georgeanna, the director of reproductive endocrinology, shared his enthusiasm but worried about protesters. Her fears were for naught, and the clinic opened with little fanfare. (Fourteen years later, when the Joneses opened one of America's first in vitro fertilization clinics in Norfolk, Virginia, protesters tried to barricade the entrance.)

Many who felt they were in the wrong body were encouraged to come to Baltimore because of the expansive view of gender proposed by John Money, a member of the Johns Hopkins team: that there were seven factors which molded gender, including not just chromosomes and genital anatomy but also behavior and sense of self. Yet at the same time, the steady flow of patients claiming to feel a gender identity different from that of their upbringing shattered one of Money's central tenets—that gender identity is malleable before the age of eighteen months. That was the thinking that prompted doctors to operate on intersex children, such as Bo Laurent in the 1950s, even switching a few baby boys with micropenis or botched circumcisions to girls before the eighteen-month deadline.

Nowadays, scientists appreciate that the development of the brain in the womb plays a crucial role in forming gender identity. Animal studies provide clues. A classic 1959 study showed that female rats born after testosterone had been injected into their pregnant mothers had ambiguous genitalia and went on to mount females the way males do. But here's what shocked the researchers: after the testosterone diminished and the females were injected with estrogen and progesterone, the rodents continued to mount other females. This was one of the first clues suggesting a hardwiring in the brain that could not be changed by hormone treatment.

However, this study, and similar ones conducted afterward, focused on mating behavior, which is vastly different from gender identity.

"People, even some scientists, talk about gender in animals all the time and maybe they have gender but we have no way of knowing. What we know about animals is their sex," said Leslie Henderson, PhD, a professor of physiology and neurobiology at the Geisel School of Medicine at Dartmouth College. Other scientists have shown that sexual behavior in animals is relevant not only to reproduction, but may also show aggression or assert dominance. Sometimes sex is barter for food. Therefore, Henderson added, we need to be cautious in making inferences about "sexual orientation" or "gender identity" from such behaviors.

Further rodent studies pointed to size differences between males and females in small regions near the hypothalamus.* In humans, a cluster of cells near the amygdala called the central nucleus of the bed nucleus of the stria terminalis is nearly twice as large in men as it is in women. The same is true of another brain region near the hypothalamus, called INAH3 (short for interstitial nucleus of the anterior hypothalamus). If or how these size differences impact gender identity is a mystery, and, as Henderson warns, size isn't everything: "It may be something else, the neurotransmitters or the number of connections, or something different."

Several studies have examined the brains of transgender people to see whether they accord with their gender identity or with their external sexual anatomy. In other words, would Mel's brain look

---

* Male rats, for instance, have a larger medial preoptic area and smaller anteroventral periventricular nucleus compared to females. But in mice, the preoptic area is the same size in males and females, which means you can't make broad extrapolations from rats to humans (or even to mice).

more male or more female? Most of these studies were small and weak, finding if anything the tiniest of correlations.

"It seems pretty clear to me that there is a durable biological component," said Dr. Joshua Safer, the director of the Transgender Medicine Research Group at Boston University. "But," he added, "we really have no clue on the specifics. The fanciest MRIs we have may show subtle differences, but give a specialist a stack of brain images and it would be impossible to know which one came from a transgender person, just as it would be tricky to tell a male from a female brain."

The feeling that one is in the wrong body is probably the result of a host of factors, including hormones, genes, and perhaps substances in the environment. And what causes transgender identity in one person may not be the same as in another.

As a child, Mel wanted to be a boy. He insisted on wearing pants from the boy's department, not skirts, nothing frilly. His mother cooperated. She even sewed custom shirts and pants for him to wear to school but demanded he wear a dress for picture day. His family assumed that Mel was a tomboy who would one day embrace a feminine side.

By the time Mel started high school, he wanted to fit in, so he grew his hair long and dressed like the girls. "Outwardly, I was very happy and gregarious and ebullient but very distant from my inner core." At the University of Arizona, he met the man who became his husband. "I felt super strongly attracted to him," he said. They carried on a long-distance relationship after graduation, and married in 1989.

In 1999, Mel started to believe that his growing sense of unhappiness was because he was a lesbian. The couple separated and divorced the following year, though they remained connected

On the left, Mel as a 6-year-old child with his older sister; on the right, Mel in high school. *Courtesy Mel Wymore.*

as friends and co-parents to their children, who were toddlers at the time. Mel's mother struggled with his coming out, which led to a tense relationship for a few years—and he wasn't happy with himself either. His relationships with women, although physically agreeable, were often filled with turmoil. "All of this time, I'm in therapy and I'm supposed to be this emancipated lesbian but I have these rocky relationships and I'm wondering why adulthood is so miserable," he said. "I started off having everything I could ever want: a career no one could complain about [as an engineer]; a perfect husband; two beautiful children. But something was really missing for me. I just kept trying to figure it out. Gender never came up in any session."

The turning point occurred during a middle school function.

As head of the diversity committee of the Parent Teacher Association, he invited the Yes Institute, which works with educators to create safe environments that empower gay, lesbian, and transgender youth, to give a talk.

"Since I was the parent coordinator, I sat at the back of the classroom. I'm sitting there and they showed a clip from Oprah Winfrey and Barbara Walters interviewing transgender kids. One of them looked just like me when I was a kid and I'm wondering if that's why I'm so unhappy."

Mel registered for a weeklong seminar at the Yes Institute's headquarters in Miami, Florida. "There were all of these trannies in the class, and a parade of people in various stages of transitioning, and it freaked me out. I had already come out and blown up my family. I'm thinking, 'Holy shit, I'm going to do this all over again.'"

Mel made his first public announcement after telling his family, with the support of his brother and two sisters. After chairing a meeting of the community board on the Upper West Side of Manhattan, attended by at least fifty people, he made an announcement: "I've had this personal discovery and that my gender has been a source of pain for me and I'll be exploring what that means, and you may see me go through some changes. I don't know where I'm going to land, but you can count on me to work hard as chair and to be open and willing to answer any questions you have regarding the transition. This is news to me as it is to you. I want your questions and I ask for your patience." He made a similar announcement at the next Parent Teacher Association meeting (he was president by then). A few people approached him and said though they were confused, they were happy for him.

Fortunately for Mel, he lived in a progressive neighborhood.

He got treatment from an endocrinologist who is considered one of the world's leading experts in the treatment of transgender people. Not all transgender individuals are so lucky. Even so, changing the person everyone thinks you are is never easy. "There's this grieving process," said Mel, like there is with any loss, even a divorce. "The person you loved, the person you imagined as part of your future, suddenly changes and that future disappears. Gender is so ingrained in our sense of self and social circles. You upset every aspect of your life when you transition. Shifting that will inherently cause grief."

Mel had a double mastectomy in 2010. Shortly afterward, he started hormone therapy. First, he took medication that blocks estrogen and accelerates menopause. Within a few months, he no longer needed the blockers because his body, like that of a post-menopausal woman, was not producing estrogen.

"That immediately made me feel better. My body was starting to feel like who I am. As I lost the estrogen feeling, I was more myself," he said.

Next, he started rubbing testosterone gel on his chest, because the gel had a slower response time than shots. "I wasn't going for the big oomph and I felt I could control my own dosing," he said. Still, the testosterone had a huge impact on his sexual desire.

"I was really taken aback by how powerful testosterone was on my sex response, just like a male in puberty. Everybody, everything was a potential sex object. I banned myself from dating. I felt I wasn't a reliable emotional being. I wanted to wait until I settled. I was literally a boy at seventeen going through puberty"—though, as Mel likes to add, he had an adult attitude with adolescent desires.

In transgender men (female to male appearance), testosterone therapy builds muscles, sprouts facial hair, boosts libido, and can also alter body odor. In transgender women (male to female appearance), estrogen works not so much because of a direct impact on the body but because it lowers testosterone. Reduced levels of testosterone shrink muscle mass and alter fat distribution, padding the hips. Some transgender women also take anti-androgens to further diminish their testosterone level.

Doctors keep an eye on other hormone side effects. Taking testosterone increases the number of blood cells, which may increase the risk of stroke and heart attack. Estrogen, according to a few studies, may increase the risk of depression. But Dr. Safer, the Boston University endocrinologist, said that, for the most part, people are so happy with the transition that any potential mental health effects from hormone treatment are outweighed.

Hormone therapy for transgender adults will not remove all the changes that occurred during puberty. Testosterone will not reduce breast size. Estrogen will not reduce the size of the Adam's apple nor turn a deep masculine voice into a higher feminine one. That's why more doctors are now willing to treat teenagers and even begin therapy when the first signs of puberty hit. The most recent Endocrine Society guidelines, published in the fall of 2017, say that some children younger than sixteen may start hormone therapy—a change from the previous guidelines, released eight years ago, which said hormones should be initiated around the age of sixteen.* But, the experts warn, data is lack-

---

* The most recent guidelines were also cosponsored by several major medical societies, including the American Association of Clinical Endocrinologists, the American Society of Andrology, the European Society for Pediatric Endocrinology, the European Society of Endocrinology, the Pediatric Endocrine Society, and the World Professional Association for Transgender Health.

ing. No large-scale trials have tracked children for many years, monitoring side effects or collecting information on how many children who report being transgender may wind up not being transgender after all.

At the same time, doctors are concerned about initiating treatment on adolescents who may decide they don't want to transition after all. Puberty blockers are reversible, so if children decide later they are not transgender, they can stop the drugs and go through a delayed puberty. Some doctors have suggested keeping young adolescents on puberty blockers as long as possible until they have the maturity to appreciate the decision, but that's complicated too. As one doctor said, it's difficult enough to have a transgender identity, but to look like a scrawny boy dwarfed by classmates who are turning into hairy, muscular young men and curvy young women can be equally traumatic. The latest guidelines offer suggestions but, as experts (and parents) know, each child has to be considered individually. A one-size-fits-all protocol will not work for everyone.

Unlike Mel and other adults who transition after having children, adolescents must consider their potential infertility. A child designated male at birth who takes anti-androgens and estrogen plummets sperm counts. Some adults have stopped taking their medication for several months to restore their fertility. Some teens may freeze sperm or eggs. But this option is not available to children who use hormone blockers at the first signs of puberty. It's impossible to retrieve sperm from prepubescent boys and mature eggs from prepubescent girls.

A father of a teen who transitioned from male to female appearance said his child considers the start of estrogen therapy as a birthday. A mother of another teen who is considering transitioning from female to male appearance said she hopes to delay testosterone therapy until her child is at least twenty years old, because she worries about the long-term impact. "Nobody can tell us the effect on the developing brain, and that scares me way more than anything else," she said. She added a caveat: If her child became depressed, then the benefits of hormone therapy would outweigh the unknown long-term risks.

Mel said the testosterone gave him a surge of confidence. When he went home to visit his family, his sister told him he was acting like a "male asshole." Was this because he was happier, which made him feel more confident? Or was it the testosterone? Mel isn't sure.

During his transition, Mel was squeamish about using bathrooms designated for men. He worried about being found out, worried that he looked too feminine or transgender, particularly because he'd have to wait for a stall. Unlike ladies' rooms, where women chat while checking themselves in the mirror, rearranging their clothing and fixing their makeup, men, Mel soon learned, aren't social urinaters. "People pee and leave."

"I was quite timid about it, and once I started it took me a while because the smell of a men's room is so oppressive and men are just so much more piggish about the condition of the restroom. I had this olfactory assault," said Mel.

Mel hopes he is one small player fostering a new understanding of gender. "The binary exists but the problem is the significance we give to it. I'd like to ghost the boxes, more blurred and less hardened in terms of the influence it has on everything that happens in your life."

Half a century ago, when Dr. Jones helped launch the Johns Hopkins Hospital's gender identity clinic, he worried that patients' expectations exceeded medical possibilities, that they expected treatment to cure their emotional angst. Today many experts fear the opposite. They worry about the impact of no treatment. More than 40 percent of transgender people have attempted suicide— ten times the national average. The vast majority of those attempts were before therapy. Mel never considered himself suicidal, but he had ten car accidents between the ages of sixteen and twenty-six. "It was kind of emblematic of not paying attention," he said, "driving really fast everywhere."

Scientists may never know what triggers the incongruence between gender identity and external genitalia, but a community of activists, investigators, and clinicians are working to figure out the best and most effective therapy to help with a safe transition in the most timely fashion possible.

15.

Insatiable: The Hypothalamus and Obesity

KAREN SNIZEK'S SON NATE was born hungry. Aggressively, insatiably hungry. As a baby he nursed all the time—except, of course, when he was sleeping, but that never lasted long. His gnawing appetite would wake him and he'd squeal for more milk. "I was so drained, I felt the life force being sucked out of me," Snizek said.

Nate didn't calm down when he started to eat real food. After each meal, his mother would pry him out of his highchair. Then he'd toddle back and cling to the chair's leg, wailing, as if he hadn't eaten in days. It was heart-wrenching for a mother to deny her child more, but she couldn't keep feeding him. Nate was born weighing "six something," she said, but soon ballooned into a rotund infant and an obese toddler.

When he was nearly two years old, Snizek had a hunch that her son wasn't just ravenous but that something was terribly wrong. No one is *that* hungry. She took him to the family practitioner.

"I'm not sure how long it took, a month or two, before she sent us to a specialist," said Snizek. That's when a blood test revealed that Nate had a rare endocrine disorder. Because of a faulty gene, the hormone that should have made him feel full wasn't doing its job. It's called POMC deficiency, short for proopiomelanocortin deficiency. Nate wasn't just a kid who liked food; he was programmed to keep eating.

The thinking about obesity today is as different as Blanche Grey, the nineteenth-century Fat Bride, is from Nate, a fat kid born in 2008. Back in Grey's day, the word "hormone" hadn't yet been invented. Massively overweight people were either gawked at in the circus or eyed quizzically by doctors who didn't know much about their condition. In the years between Grey's death and Nate's birth, the field of endocrinology was born and flourished. Investigators cleared paths through a tangled forest of data. Nate's specialist spotted the hormone defect right down to the region of his missing genes—2p23.2. This crucial snippet of information, along with other recent findings, is paving the way toward new drugs that may quench the cravings of people like Nate and help the rest of us curb our appetite. On a more fundamental level, emerging endocrine research is providing a fresh perspective into the biological basis of one of our most primal urges, the drive—the hormonal drive, that is—that compels us to eat.

Early on, most physiological studies of weight gain focused on energy, trying to figure out why some people burned calories faster than others. Overeating was the domain of psychologists, who dealt with emotions. Hunger wasn't considered hormonal until the 1950s. That's when scientists began studying fat rats. Some of them were born that way; others were force-fed. (Rats don't vomit, which

makes enforced weight gain easier and less messy.)* Also around
that time, researchers began to appreciate the immense influence
of the hypothalamus, an almond-sized brain gland which keeps the
body in check with a vast array of hormones. The hypothalamus is
home to hormones that control, among other things, body temperature, stress, and reproduction. Researchers wondered if it might
control appetite, too. The evidence suggested as much; removing
a chunk of a rat's hypothalamus drove the rodent into a feeding
frenzy. Perhaps a crucial hormone had been removed.

In a stunningly simple yet quirky experiment conducted in
1958, George R. Hervey, a scientist at the University of Cambridge, joined two rats by peeling away their skin and sewing them
together. It was as if he had created conjoined twins. Blood circulated through both. Hervey's hunch was that if the rats shared
blood, they would share hormones too. He plucked out the hypothalamus of one rat, triggering insatiable hunger and, therefore,
obesity. He speculated that the mucked-up hormones of the first
rat might float across to the other rat, causing it to overeat too, but
just the opposite happened. The second rat refused food. Hervey
ran the experiment with several rat pairs, and the result was always
the same. "Even when [food] was offered by hand," he wrote, "they
did not eat but simply looked away." The rats who had received
brain surgery got really fat, while their partners starved themselves to death.

Simply put, consuming calories fires up a chemical reaction

---

* Rats don't burp, either, according to ratbehavior.org. That's because they have neither the
musculature in the gut nor the brain-body coordination to push food back up. Vomiting is
useful for clearing toxic food. Since they cannot do it, rats are picky eaters, taking a tiny taste
first in order to avoid ingesting something dangerous. I find this hard to believe, as the rats
on my New York City block seem amenable to anything tossed outside, including rat poison.

that shuts down appetite, Hervey explained in an article in the 1959 *Journal of Physiology*. He concluded that his study lent support to a burgeoning theory of hormonal "negative feedback." The level of one hormone in the body goes up, which signals another hormone that dampens it. This is the body's way of maintaining balance, what doctors call homeostasis. That's how the menstrual cycle works with seesawing estrogen and progesterone; that's how the pancreas controls sugar levels. Hervey figured that when the fat rat ate, it boosted an "I feel full" hormone. The rodents kept eating because of their defective brains, but when that hormone circulated into the thin rat's bloodstream, its signal was received. Though the thin rat had not eaten anything at all, the hormone served as a stop sign, alerting it—albeit falsely—that it had eaten enough.

These early British rat studies, along with further ones conducted in the U.S., launched a hunt for the elusive substance. In 1949, George Snell, working at the Jackson Laboratory in Bar Harbor, Maine, found a strain of mice that were triple the weight of other mice. They ate ferociously. He named them "ob" mice, for obesity. Ten years later, Douglas Coleman, also at the Jackson Laboratory, found another strain of mice that were fat and ate a lot, and also had diabetes. He called them "db" mice, for diabetes. Coleman did a surgical Siamese-twin study similar to Hervey's. He suspected that these mutant mice must lack a "satiety factor," but its identity remained a mystery.

In the 1970s, Rosalyn Yalow—the Nobel Prize–winning scientist who created radioimmunoassay—proposed that cholecystokinin, or CCK, which is secreted by both the gut and the brain, was the sought-after hormone. Much to her chagrin, her protégé Bruce Schneider proved her wrong. We now know that CCK is

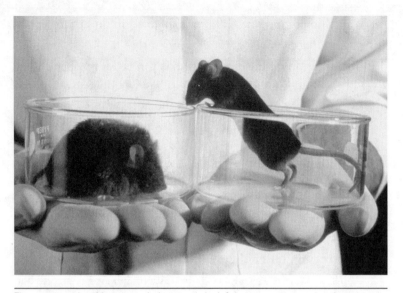

The mouse on the right is normal; the one on the left has a genetic mutation that triggers a leptin deficiency, leading to voracious eating. *Remi Benali/Gamma-rapho/Getty Images.*

released during eating and stimulates digestion; it has to do with hunger, but it's not the "I feel full" hormone. The search gained momentum in the early 1980s, when scientists had better tools to spot genes that make hormones. Even then, it took another decade or so to pinpoint the precise gene. (Gene hunting is like a scavenger hunt—first scientists look for the general region, then there's a rigorous process of homing in closer and closer as they gather clues and spot the prize.)

In 1994, inspired by Coleman's work and tapping into new gene-tracking techniques, a team led by Dr. Jeffrey Friedman at Rockefeller University found the gene that produced the hormone. It wasn't as simple as had been assumed: it was not an "I feel full" hormone, but more of a weight-controlling hormone. It controls appetite over the long haul, establishing the set points when hunger strikes and when the feeling of fullness occurs. If it does not function properly, the feeling of fullness never comes.

The hormone was named leptin, from the Greek *leptos*, "thin." Of all the unlikely hiding spots, leptin was found in a fat cell. That was a real shocker, because it suggested that fat cells were not just greasy blobs but endocrine organs like the ovary and the testes and all the rest.

"We didn't appreciate that in addition to being a fuel depot, they release all sorts of molecules," said Rudy Leibel, a Columbia University professor of pediatrics and director of the Division of Molecular Genetics and the Naomi Berrie Diabetes Center. Leibel had worked with Friedman at Rockefeller and played a crucial role in the steps leading up to the cloning of the gene.

But it was the hormone discovery itself—not its fatty habitat—that sparked a sensation in the greater scientific community and among dieters worldwide, lending credence to speculation that some of us are plump because of chemistry, not flagging willpower. "Genes, not Greed, Make You Fat" was the headline in the *Independent* of London. The *New York Times* led its report with, "Lending mighty support to the theory that obese people are not made but, rather, born that way."

Doctors now consider leptin the hormone equivalent of an alarm system, alerting the body when it is starving. Sinking energy stores trip a switch that fuels hunger. Consumption cools hunger, and the leptin alarm goes back into resting mode. When no food is ingested, leptin remains at a dangerously low level, disturbing other hypothalamic hormones: slowing reproduction and metabolism, withering the immune system. All of these biological processes (making babies, fending off germs, heating the body, and so forth) consume energy, explained Friedman, "so when leptin is low, you dampen down a whole set of responses that increase energy expenditure."

That's why women who starve themselves stop menstruating, become infertile, and are prone to illness. Other hormones—perhaps sparked by leptin—are thrown off kilter, too, prompting excess hair growth on arms and fragile, brittle bones, both of which are common among anorexics. Doctors have known about these dangers for years, but only recently has leptin been identified as the linchpin that triggers the turmoil.

Leptin is also a reason why so many of us have trouble keeping weight off. It is the hormonal explanation for the old set point theory. Most people tend to have a weight that is "normal," or feels right for them, even though it may not be the fashionable rail-thin physique. When we eat less and shrink our fat stores, leptin dips too, prompting cravings to eat our way back to our curvier starting point. The good news is that it works both ways. When we are left to our own inner chemical devices, hunger tends to diminish after a binge, bringing us easily back to where we started. Sustainable weight loss is possible with slow, gradual diets that adjust the set point, perhaps by resetting leptin levels.

Given these findings, it seemed likely—or wishful—that leptin would stave off the hunger people feel when dieting, helping them keep off the pounds. Unfortunately for weight-watchers and for the drug-makers hoping for a blockbuster, leptin shots only work for the extremely rare folks born without any leptin at all. That's probably because we eat for so many reasons, often when we're not even hungry. Also, there are many more hormones that control hunger, fullness, and how quickly calories are burned.

Nate does not lack leptin, but has a defect in the leptin receptor in his hypothalamus. The outcome is the same: he feels relentlessly hungry. The difference is that all the leptin in the world wouldn't help him. Giving him leptin would be like wrapping duct tape around the wrong spot in a leaky hose.

Because of the defect in his leptin cascade, Nate has a host of endocrine issues. In Blanche Grey's day, children born with Nate's defect didn't survive infancy. He takes cortisone pills three times a day for his adrenal glands, and thyroid once a day for his thyroid gland. Without the leptin pathway working, he won't go through puberty on his own, so he will need sex hormones when the time comes.

Thanks to research into people like Nate, scientists have made huge headway into the physiology of hunger. But we are in the very earliest days. Clues emerging now will provide the scaffolding for future studies of appetite and energy. Endocrinologists are joining forces with infectious disease specialists, immunologists, neuroscientists, and even environmental specialists. For instance, the trillions of germs living in the digestive tract—the microbiome—spew out their own chemicals, which may alter the way hormones sway appetite or the way the body burns calories. That is, some germs may give us a tendency to gain weight, while others may give us a tendency to lose it. Some studies suggest that antibiotics make people more likely to gain weight because they kill off bacteria in a way that promotes obesity. Others have shown just the opposite. It's too early to draw conclusions, so we can't count on a probiotic drink to replenish the "good" germs. There's a chance that insights about the gut will be linked to insights about leptin, providing a thorough understanding of hunger. Other studies focus on air pollution and industrial chemicals and pesticides that make their way into water and the food supply. Some of these toxins, like computer hackers, may switch on hormones, foiling the system. Still, scientists are a long way from confirming their hunches.

Even weight loss surgery, once thought to work by reducing the amount of space for food, is now considered to work because of the way it alters hunger hormones. Maybe scientists can devise a weight loss drug that would be preferable to a risky operation—but that begs the question whether a drug with side effects is safer than surgery.

At Columbia University, Rudy Leibel, the pioneering obesity researcher, is trying to understand hunger signals at the cellular level, conducting experiments to understand how multiple hormones send hunger messages between brain cells. What's really needed, he said, are sophisticated studies of humans, using better imaging machines than are available now. "We're not there yet in terms of the control of food intake," said Leibel. "We will be, and then we will have a much better idea of how these endocrine systems act and how they act on the brain and the [gastrointestinal] tract."

Though we can't dismiss the very real fact that some of us are eating too much, our bodies may be less efficient calorie burners, or hungrier machines, because of the way our hormones are functioning. Laboratory animals are fatter than they used to be, and they're given the same kibble as before. Are they hungrier? Not burning off the calories as easily? No one knows, but it's another hint that something is going on, something that may be altering hormones and feeding the obesity epidemic. Or maybe we are supersizing ourselves into hormonal disarray. Studying appetite in humans is tricky because researchers aren't starting with a clean slate. It's the old chicken-and-egg problem: Were you born with a propensity to gain weight because of the brew of chemicals inside you, or did you acquire a fat-prone physiology because of your diet? Did your mother's eating habits during pregnancy, or exposure to certain chemicals, influence the way you handle extra calories or the temptations of junk food? Are we drowning in a swamp of contaminants that are obesogenic? (That's a new word to describe things that make us fat.) Or is it that the way we live—every holiday, every social occasion—is so food-centric?

Nate is now eight years old. He is round with short, stocky legs. His body mass index—a measure of weight to height—is well above the obesity level. He's generally a cheerful kid, thanks to his devoted mother who goes all out to keep him upbeat and to distract his attention from food. It's a full-time job—so all-consuming that he is home-schooled so that she can regiment his eating to carefully calorie-measured, frequent meals, and keep an eye on him at all times to make sure he doesn't cheat. The two of them live in a gated community in New Smyrna, Florida, a beach town about an hour outside Orlando.

Three friends with issues similar to Nate's, whom Snizek met on Facebook, are taking an experimental drug that switches on the leptin receptor. But Nate is still a decade away from qualifying for the study: volunteers must be at least eighteen years old. One volunteer disclosed to Snizek that he's lost 25 pounds in four months, and feels full for the first time in his life. Food, he says, has never tasted so good.

Nate's mother is, naturally, frustrated. "Just give me the damn medicine, yesterday," she said.

She's eager—anxious—to get Nate into any study she can, whether it's a trial of a new appetite suppressant or one studying his host of hormones gone awry. "I think his body is holding answers," she said. When he was four, Nate spent a week at the

National Institutes of Health in Bethesda, Maryland. She thinks he provided more information for the doctors than they were able to provide for her, but she felt she did the right thing for the future of endocrinology.

Research into obesity is more than merely a study of weight gain. It is on the forefront of endocrinology because it connects cells to behavior in ways that the twentieth-century pioneers of hormone science longed to do, but never could. "We humans carry with us the deep-seated wish to have control," said Friedman, the discoverer of the leptin gene. "For obesity, there is an illusion of control because you can lose weight if you stop eating, but that ignores the fact that there is a basic drive that will push you to eat in the same way there is a basic drive that pushes you to drink fluids or have sex or do lots of other things. I don't think on the whole that humans have come to grips entirely with how powerful our basic drives are, how difficult it is to use conscious means to control them." How our hormones, that is, control just about everything.

Epilogue

IN 1921, NEARLY TWENTY YEARS into hormone research and nearly five years after the creation of the Association for the Study of Internal Secretions, Harvey Cushing, the brain-collecting neurosurgeon, thought it was a good time to reflect on the status of this inchoate field. Recent discoveries into the pituitary, he said, had led to a "deluge of papers, and soon after that, the endocrinologist came into being." Cushing was famous for a lot of things, modesty not among them. His introduction was a nod to his own research; he was giving himself credit as the father of the field. Or, as he went on to tell it, the captain of the ship. "Some unquestionably follow the lure of discovery; some are earnest colonizers; some have the spirit of missionaries and would spread the gospel; some are attracted by gain and are running full sail before the trade wind," he said. "The impulse which launched us in such amazing numbers on our several missions some future historian may have to tell."

Cushing hailed the triumphs of his dedicated colleagues, such as the use of thyroid hormone for those with defective glands and the discovery that faulty adrenal glands spur Addison's disease. He accepted that there would be missteps along the way. He probably wasn't surprised that, a few years after his talk, well-intentioned but mistaken physicians provided expert testimony for two murderers, claiming that their pineal glands made them do it. Cushing wasn't around, but he could have predicted the almost religious zeal that launched the global pituitary collection, eking out droplets of growth hormone for short kids. He was a stickler for precision—that's what made him one of the best brain surgeons of his time—and might even have spotted the sloppy lab technique that led to contaminated batches.

Cushing was familiar with and infuriated by hucksters peddling cockamamie concoctions touting all kinds of youth-enhancing, libido-boosting hormone remedies, so if his spirit were floating about today, he wouldn't be surprised to see the Low-T commercials or the oxytocin love drugs. But he might not have foreseen the advent of radioimmunoassay, which transformed hormone medicine from guesswork to precision, measuring hormones down to the billionth of a gram. And he might not have predicted that germs in the gut and the lowly fat cell would be considered hormone producers, alongside the esteemed pituitary.

The story of hormones is the story of interconnectedness. It began with an examination of the glands within us, focusing solely on those clusters of cells dedicated to making hormones. Cushing was amazed by the links between belly and brain, the hormones ricocheting this way and that. Today, we are beginning to appreciate that each of us is one small pond within a vast ocean of hormone-changing chemicals.

"We find ourselves embarked on a fog-bound and poorly charted sea of endocrinology," Cushing told doctors nearly a hundred years ago at a meeting of the American Medical Association in Boston. "It's easy to lose our bearings for we have, most of us, little knowledge of the seafaring and only a vague idea of our destinations."

The twenty-first-century voyagers—the kinds of investigators that Cushing dubbed "earnest colonizers"—have gained a sharper focus. Not crystal clear, but less murky all the same. They are exploring technologies that will help them spot hormone-making genes, and help them visualize the microscopic impact of these ever-so-tiny and ever-so-potent chemicals. As for us, if history is our guide, we have become more discerning consumers of drugs and information. We've been inoculated with a healthy skepticism to help us navigate the flood of hype and hope and to grasp the hardcore science that will keep this ship on course. As we venture forth, we are sure to gain a better appreciation of the chemical tugs that make us yearning, moody, hungry people. In other words: the chemistry of being human.

Acknowledgments

The process of writing this book has been a lot like the way hormones work. I don't mean triggering mood swings—rather, that hormones rarely act alone. They rely on other hormones to guide them, to nudge them, to signal when it's time to calm down. Same for me. I relied on experts, friends, and family who guided me, nudged me, and signaled when I needed to chill out.

Many specialists who provided me with information are listed in the notes to the relevant chapters. A few doctors were on call for me 24/7. Thank you, Dr. Mary Jane Minkin, clinical professor of obstetrics and gynecology at Yale University, my go-to menopause person; Dr. Alexander Pastuszak, assistant professor in male reproductive medicine and surgery at Baylor College of Medicine, my testosterone guy; Leslie Henderson, PhD, professor of physiology and neurobiology, Geisel School of Medicine at Dartmouth College; and Dr. Joshua Safer, director of the Transgender Medicine and Surgery Team, Boston Medical Center, whose expertise helped me fine-tune my word choices.

As always, thanks to all the librarians who have tracked down obscure material: Arlene Shaner, Historical Collections Librarian at the New York Academy of Medicine; Melissa Grafe, head of the medical history library at Yale's Cushing/Whitney Medical Library; and Stephen E. Novak, head of archives and special collections at the Augustus C. Long Health Sciences Library at Columbia University Medical Center. The New York Academy of Medicine's Walter Linton and Tsz Chen Lee retrieved and emailed scientific articles not available online.

It's always so nice to know that there are generous researchers and writers willing to share their sources and material. Thank you to John Hoberman, historian, writer, and professor of Germanic studies at the University of Texas at Austin; Emily Green, an intrepid reporter who covered the growth hormone story when it first broke; and Jonathan Eig, author of *The Pill*, who shared taped interviews with the birth control pill discoverers.

Howard and Georgeanna Jones's children, Larry Jones, Dr. Howard Jones III, and Dr. Georgeanna Klingensmith, were so kind as to invite me into their homes and share information about their parents. Elanna Yalow and Ben Yalow spoke with me about their mother, Roslyn Yalow.

Thanks from the bottom of my heart to my wonderful friends who provided me with honest feedback and found time to read and reread (and reread) drafts during their own hectic schedules. My New Haven crew: Anna Reisman, Lisa Sanders, John Dillon, and our newest member Marjorie Rosenthal. I look forward to our continued three-hour Cedarhurst Coffees. My New York squad: Judith Matloff, Katie Orenstein, and Abby Ellin (the title queen) for detailed comments, for responding to my emails at all hours, for spotting errors. Thank you to Sheri Fink, Elyse Lackie, Annabella Hochschild, and Jessica Friedman for wise comments that shaped my stories. I always enjoyed my Upper West Side evenings with fellow writers Marie Lee, Jon Reiner, and Alice Cohen. I'm also grateful to the brilliant fellows and co-mentors at the OpEd Project who are always willing to share their expertise. To Catherine McGeoch, and Jessica Pevner for close reading and creativity.

And of course the Invisible Institute, whose increasingly visible members have provided me with advice and encouragement at our monthly dinners.

The opportunity to talk about my research in front of real audiences encouraged me to hone my thoughts—and get words on paper. Thanks to the New York Academy of Medicine, particularly Lisa O'Sullivan, vice president and director of the library and Center for the History of Medicine and Public Health, and Emily Miranker, the events and projects coordinator. Thanks to Dr. Jay Baruch for inviting me to speak at Brown University's Cogut Institute for the Humanities. I was a medical student when the Humanities in Medicine Program at Yale University launched and now I'm happy to serve on its board for the director, Dr. Anna Reisman (the Anna of my writers' group). She's given me the opportunity to speak, including once when she wanted me to talk about a chapter I had barely finished. Nothing like tight deadlines to get me going.

For moral support and for *never* telling me to stop talking about hormone history and the writing process when I know they really wanted to: Marguerite Holloway, Cathy Shufro, Harriet Washington, Lauren Sandler, Laurie Niehoff, Lizzie Reis, Joanna Radin, Wendy Paris, Alice Tisch, Tommy Tisch, Doug Kagan, Adina Kagan, and Jane Bordiere. Dr. Tom Duffy, professor emeritus of medicine, Yale University, regaled me with stories of his days at Johns Hopkins. Dr. Manju Prasad, director of endocrine, head and neck pathology at Yale University, read drafts and also provided wise insights into both writing and medicine. Johanna Ramos-Boyer and Virginia Shurgar Hassell for motivating me. Mark Schoenberg and Risa Alberts for chauffeuring me around Baltimore in addition to being great listeners. Jessica Baldwin, my friend in London, dashed to Battersea Park to hunt down a hidden statue because I needed one more image ASAP. Dr. Chuck Sklar, pediatric endocrinologist and director of the long-term follow-up program at Memorial Sloan Kettering Cancer Center, and Dr. Myron Genel, professor emeritus of pediatrics, Yale University, put big issues in endocrinology into focus for me.

I learn so much from my students. Thank you, Tali Woodward, for inviting me into the Columbia Journalism School family where my students continue to be a source of inspiration. To Andrew Ehrgood, for providing me with the opportunity to work with undergraduates; my English 121 classroom conversations about nut grafs, omission, structure, ledes, and jettisoning medical jargon help my writing. The classroom enthusiasm is contagious. Our conversations energize me.

I am forever indebted to Joy Harris, a writer's dream agent. She's also a good friend and a good person. My team at W. W. Norton supported me through this process. I'm so thankful that Jill Bialosky (who is also a poet, novelist, and memoirist) has been my editor for both of my books. She helped me develop my voice and steered *Aroused* in the right direction when I started to stray. Thanks to the entire W. W. Norton team: Amy Medeiros, project editor; Ingsu Liu, art director for the cover; Lauren Abbate, production manager. Drew Elizabeth Weitman is an editorial assistant extraordinaire. I showered/bombarded her with questions and she responded cheerfully and quickly to every single one. The wickedly smart copyeditor Allegra Huston fine-tuned my manuscript.

You don't really appreciate mothers until you are one. During the course of writing this book, I cut off my mother a lot when she called while I was writing. But I expected her full attention when I reached out to her regardless of her busy schedule. Thanks, Ma. My late father, Dr. Robert V. P. Hutter, cared a lot about words, about medical data, about scientific truths, and about communicating medicine accurately, honestly, and with empathy. I like to think he would have been proud. My brother, Andrew, and my sister, Edie, have always believed in my abilities when my confidence floundered.

My children Jack, Joey, Martha, and Eliza are my everything. They even provided unsolicited editing advice, such as insisting that my hormone/*Aroused* puns be deleted from this acknowledgment. And finally, of course, Stuart—it's the chemistry.

Notes

1. THE FAT BRIDE

Details of the life and death of Blanche Grey are taken from: "Trying to Steal the Fat Bride: Resurrectionists Twice Baffled in Attempts to Rob The Grave," *New York Times*, October 20, 1883; "The Fat Girl's Funeral: Her Remains Deposited in a Capacious Grave at Mt. Olivet," *Baltimore Sun*, October 29, 1883; "More than a Better Half," *New York Times*, September 26, 1883; "The Fattest of Brides Dead," *Baltimore Sun*, October 27, 1883; "Her Fat Killed Her," *Chicago Daily Tribune*, October 27, 1883; "Poor Moses: How the Late Fat Girl's Husband was Scared," *San Francisco Chronicle*, November 19, 1883; "Sudden Death of a 'Fat Woman,'" *Weekly Irish Times*, November 17, 1883; and "A Ponderous Bride," *Baltimore Sun*, October 1, 1883. An overview of the early years of endocrinology is given in V. C. Medvei's thorough *A History of Endocrinology* (Lancaster, U.K.: MTP Press), 1984.

1 **body snatchers:** "Trying to Steal the Fat Bride: Resurrectionists Twice Baffled in Attempts to Rob the Grave," *New York Times*, October 20, 1883.
3 **voyeuristic shows:** Robert Bogdan, *Freak Show: Presenting Human Oddities for Amusement and Profit* (Chicago: University of Chicago Press, 1998); Rachel Adams, *Sideshow U.S.A.: Freaks and the American Cultural Imagination* (Chicago: University of Chicago Press, 2001).
3 **stoked the fascination of an eclectic crew:** Aimee Medeiros, *Heightened Expectations* (Tuscaloosa: University of Alabama Press, 2016).
4 **an autopsy . . . revealed a tumor:** Fielding H. Garrison, "Ductless Glands,

Internal Secretions and Hormonic Equilibrium," *Popular Science Monthly* 85, no. 36 (December 1914): 531–40.

4 **developmentally delayed ten-year-old:** J. Lindholm and P. Laurberg, "Hypothyroidism and Thyroid Substitution: Historical Aspects," *Journal of Thyroid Research* 2011 (March 2011): 1–10.

4 **museum . . . Monroe Hotel:** Steve Cuozzo, "$wells Take Bowery," *New York Post*, December 26, 2012.

4 **"adipose monstrosity":** "More Than a Better Half," *New York Times*, September 26, 1883.

5 **"freaks' boarding house":** "The Fat Bride," *Australian Town and Country Journal*, January 12, 1884.

6 **realized she was dead:** "The Fat Bride," *Manawatu Times*, January 28, 1884, available at: http://paperspast.natlib.govt.nz/cgi-bin/paperspast?a=d&d=MT18840128.2.20.

6 **"The crowd on the pavement":** "The Fat Girl's Funeral: Her Remains Deposited in a Capacious Grave at Mt. Olivet," *Baltimore Sun*, October 29, 1883.

9 **"The physician without physiology":** Roy Porter, *The Greatest Benefit to Mankind: A Medical History of Humanity* (New York: W. W. Norton, 1997), 305.

11 **"They crowed lustily":** Homer P. Rush, "A Biographical Sketch of Arnold Adolf Berthold: An Early Experimenter with Ductless Glands," *Annals of Medical History* 1 (1929): 208–14; Arnold Adolph Berthold, "The Transplantation of Testes," translated by D. P. Quiring, *Bulletin of the History of Medicine* 16, no. 4 (1944): 399–401.

11 **He published his insights:** Rush, "A Biographical Sketch."

11 **as if Columbus discovered America:** Albert Q. Maisel, *The Hormone Quest* (New York: Random House, 1965).

12 **Thomas Blizard Curling:** Lindholm and Laurberg, "Hypothyroidism and Thyroid Substitution."

12 **Thomas Addison:** Henry Dale, "Thomas Addison: Pioneer of Endocrinology," *British Medical Journal* 2, no. 4623 (1949): 347–52.

12 **George Oliver:** Ibid.

12 **named "adrenaline":** Michael J. Aminoff, *Brown-Séquard: An Improbable Genius Who Transformed Medicine* (New York: Oxford University Press, 2011); Porter, *The Greatest Gift to Mankind*, 564; John Henderson, *A Life of Ernest Starling* (New York: Oxford University Press, 2005).

2. HORMONES . . . AS WE MAY CALL THEM

Details of the Brown Dog Affair are drawn from Peter Mason, *The Brown Dog Affair: The Story of a Monument that Divided a Nation* (London: Two Sevens, 1997) and Henderson, *A Life of Ernest Starling*, as well as Hilda Kean, "An

Exploration of the Sculptures of Greyfriars Bobby, Edinburgh, Scotland, and the Brown Dog, Battersea, South London, England," *Journal of Human–Animal Studies* 11, no. 4 (2003): 353–73; J. H. Baron, "The Brown Dog of University College," *British Medical Journal* 2, no. 4991 (1956): 547–48; David Grimm, *Citizen Canine: Our Evolving Relationship with Cats and Dogs* (New York: Public Affairs, 2014); and Coral Lansbury, *The Old Brown Dog: Women, Workers, and Vivisectionists in Edwardian England* (Madison: University of Wisconsin Press, 1985). Details of endocrinology in the early 1900s are drawn from Medvei, *A History of Endocrinology*; Merriley Elaine Borell, "Origins of the Hormone Concept: Internal Secretions and the Physiological Research 1895–1905," PhD thesis in the history of science, Yale University, 1976.

15 "A place to avoid": Mason, *The Brown Dog Affair*, 25.

16 "centuries of angst": Grimm, *Citizen Canine*, 48.

17 "There cannot be one standard": Mason, *The Brown Dog Affair*, 45.

17 "It's only them brown doggers": Ibid. 48.

18 declined an invitation to be knighted: Diana Long Hall, "The Critic and the Advocate: Contrasting British Views on the State of Endocrinology in the Early 1920s," *Journal of the History of Biology* 9, no. 2 (1976): 269–85.

18 Bayliss married Starling's sister: Henderson, *A Life of Ernest Starling*.

18 Starling married into money: Rom Harré, *Pavlov's Dogs and Schrödinger's Cat: Scenes from the Living Laboratory* (Oxford: Oxford University Press, 2009).

20 The point was to put the mixture: Ibid.

20 "chemical reflex": Irvin Modlin and Mark Kidd, "Ernest Starling and the Discovery of Secretin," *Journal of Clinical Gastroenterology* 32, no. 3 (2001): 187–92.

20 announced their new ideas: Barry H. Hirst, "Secretin and the Exposition of Hormonal Control," *Journal of Physiology* 560, no. 2 (2004): 339.

21 "therefore rather sceptical": W. M. Bayliss and Ernest H. Starling, "Preliminary Communication on the Causation of the So-Called 'Peripheral Reflex Secretion' of the Pancreas," *Lancet* 159, no. 4099 (1902): 813.

21 "Of course, they are right": Modlin and Kidd, "Ernest Starling and the Discovery of Secretin."

21 "The secretion must therefore": W. M. Bayliss and Ernest H. Starling, "On the Causation of the so-called 'Peripheral Reflex Secretion' of the Pancreas (Preliminary Communication)," *Proceedings of the Royal Society* B69 (1902): 352–53.

22 suggest that postmenopausal declines: Jukka H. Meurman, Laura Tarkkila, Aila Tiitinen, "The Menopause and Oral Health," *Matiritas* 63, no. 1 (2009): 56–62.

22 created a discipline: Modlin and Kidd, "Ernest Starling and the Discovery of Secretin."

22 Scientists know that secretin: Hirst, "Secretin and the Exposition of Hormonal Control."

22 secretin also regulates electrolytes: Jessica Y. S. Chu et al., "Secretin as a neurohypophysial factor regulating body water homeostasis," *PNAS* 106, no. 37 (2009): 15961–66.

22 "A chemical sympathy": Bayliss and Starling, "On the Causation."

25 "twofold": Lizzy Lind af Hageby and Leisa Katherina Schartau, *Shambles of Science: Extracts from the Diary of Two Students of Physiology* (London: Ernest Bell, 1903).

25 "His master may have lost him": Ibid.

26 "cowardly, immoral and detestable": Mason, *The Brown Dog Affair*, 11.

27 Bayliss, who shunned publicity: Details of the trial are taken from "Bayliss v. Coleridge," *British Medical Journal* 2, no. 2237 (1903): 1298–1300; "Bayliss v. Coleridge (Continued)," *British Medical Journal* 2, no. 2238 (1903): 1361–71; and "Was It Torture? The Ladies and the Dogs, Doctors and the Experiments," *Daily News*, November 18, 1903.

28 "Here is an animal": "He Liveth Best Who Loveth Best, All Things Both Great and Small," *Daily News*, November 19, 1903.

28 sneaky and reprehensible: Mason, *The Brown Dog Affair*, 19–20.

28 "bringing vile charges": "The Vivisection Case," *Globe and Traveller*, November 18, 1903.

28 four weekly lectures: Ernest H. Starling, *The Croonian Lectures on the Chemical Correlation of the Functions of the Body*, Royal College of Physicians, 1905, available at: https://archive.org/details/b2497626x.

28 "These chemical messengers": Ibid.

29 turned to two friends: Medvei, *A History of Endocrinology*, 27; Hirst, "Secretin and the Exposition of Hormonal Control."

29 "autocoid": Sir Humphry Rolleston, "The History of Endocrinology," *British Medical Journal* 1, no. 3984 (1937): 1033–36.

30 "chalone": Ibid.

30 He avoided mentioning the testes: Henderson, *A Life of Ernest Starling*.

31 "An extended knowledge": Starling, *The Croonian Lectures*, 35.

31 "seems almost like a fairy tale": Henderson, *A Life of Ernest Starling*, 153.

31 "outrageous," "mute testimony": "Battersea Has a Brown Dog," editorial, *New York Times*, January 8, 1908.

32 On March 10: Marjorie F. M. Martin, "The Brown Dog of University College," *British Medical Journal* 2, no. 4993 (1956): 661.

32 "the statue or anything of its likeness": "Battersea Loses Famous Dog Statue," *New York Times*, March 13, 1910.

32 A second *Brown Dog* memorial: Hilda Kean, "The 'Smooth Cool Men of Science': The Feminist and Social Response to Vivisection," *History Workshop Journal*, no. 40 (1995): 16–38.

3. PICKLED BRAINS

Details of Harvey Cushing's life are drawn from Michael Bliss, *Harvey Cushing: A Life in Surgery* (New York: Oxford University Press, 2005), and Aaron Cohen-Gadol and Dennis D. Spencer, *The Legacy of Harvey Cushing* (New York: Thieme Medical Publishers, 2007), which includes images from Cushing's operations (the photos are on display at Yale). I also scoured Cushing's correspondence, in the Harvey Williams Cushing Papers, MS 160, Manuscripts and Archives, Sterling Memorial Library, Yale University. I conducted interviews with Dr. Dennis Spencer, Harvey and Kate Cushing Professor of Neurosurgery, Yale University; Dr. Christopher John Wahl, orthopedic surgeon at Orthopedic Physicians Associates, Seattle, WA: Dr. Tara Bruce, obstetrician gynecologist, Houston, TX; Dr. Gil Solitaire, retired neuropathologist; and Terry Dagradi, photographer and coordinator at the Cushing Center, Yale University.

38 **"In the first decade":** Bliss, *Harvey Cushing*, 166.
38 **he boasted a mortality rate:** Ibid., 274.
38 **"Whatever approach":** Dr. Dennis Spencer, author interview.
40 **"The Chief's first and only true love":** Bliss, *Harvey Cushing*, 481.
40 **first human-to-human pituitary transplant:** Courtney Pendleton et al., "Harvey Cushing's Attempt at the First Human Pituitary Transplantation," *Nature Reviews Endocrinology* 6, no. 1 (2010): 48–52.
40 **Newspapers heralded it as a scientific breakthrough:** "Part of Brain Replaced: That of Dead Infant Put in Cincinnati Man's Head, First of its Kind," *Baltimore Sun*, March 26, 1912; "Given Baby's Brain," *Washington Post*, March 26, 1912; "Brain of Still-Born Infant Used to Restore Man's Brain," *Atlanta Constitution*, March 27, 1912.
41 **gave the dogs a piece of pituitary:** Harvey Cushing, "Medical Classic: The Functions of the Pituitary Body," *American Journal of the Medical Sciences* 281, no. 2 (1981): 70–78.
41 **He measured the skull:** Harvey Cushing, "The Basophil Adenomas of the Pituitary Body and Their Clinical Manifestations (Pituitary Basophilism)," *Bulletin of the Johns Hopkins Hospital* 1, no. 3 (1932): 137–83; Harvey Cushing, *The Pituitary and Its Disorders: Clinical States Produced by Disorders of the Hypophysis Cerebri* (Philadelphia: J. B. Lippincott, 1912).
42 **slipped the undertaker fifty dollars:** Wouter W. de Herder, "Acromegalic Gigantism, Physicians and Body Snatching. Past or Present?" *Pituitary* 15 (2012): 312–18.
42 **case after case of men and women:** Cushing, *The Pituitary and Its Disorders*.
44 **article entitled "Uglies":** "Uglies," *Time*, May 2, 1927.
44 **"This unfortunate woman":** John F. Fulton, *Harvey Cushing: A Biography* (Springfield, IL: Charles C. Thomas, 1946), 304.

46 **Now we know he may have been right:** "Pituitary Tumors Treatment (PDQ) Patient Version," National Cancer Institute, 2016, http://www.cancer.gov/types/pituitary/patient/pituitary-treatment-pdq.

46 **A doctor at the Mayo Clinic:** V. C. Medvei, "The History of Cushing's Disease: A Controversial Tale," *Journal of the Royal Society of Medicine* 84, no. 6 (1991): 363–66.

47 **Anti-Pituitary Tumor Club:** Ibid.

47 **"temptation of impressionistic speculation":** Cushing, "The Basophil Adenomas of the Pituitary Body."

47 **10,48 words a day:** Fulton, *Harvey Cushing.*

48 **jumble of jarred brains:** Dr. Gil Solitaire, author interview.

49 **"I think a few people":** Dr. Christopher John Wahl, author interview.

50 **Wahl would write a thesis:** Christopher John Wahl, "The Harvey Cushing Brain Tumor Registry: Changing Scientific and Philosophic Paradigms and the Study of the Preservation of Archives," medical school thesis in neurosurgery, Yale University, 1996.

53 **In the summer of 2017:** personal interviews with Dr. Maya Lodish and Dr. Cynthia Tsay, March 1, 2018. Also Cynthia Tsay et al., "Harvey Cushing Treated the First Known Patient with Carney Complex," *Journal of the Endocrine Society* 1, no. 10 (2017): 1312–21.

4. KILLER HORMONES

Details of the murder and the trial are drawn from Simon Baatz, *For the Thrill of It: Leopold, Loeb, and the Murder that Shocked Chicago* (New York: Harper, 2008); Hal Higdon, *Leopold and Loeb: The Crime of the Century* (Champaign, IL: University of Illinois Press, 1999), and excerpts of the trial proceedings available at Famous Trials, a website of the University of Missouri–Kansas City School of Law (http://famous-trials.com/leopoldandloeb) and in the archives of Northwestern University Library (http://exhibits.library.northwestern.edu/archives/exhibits/leoloeb/index.html). An overview of endocrinology in the 1920s was provided by Julia Ellen Rechter, "The Glands of Destiny: A History of Popular, Medical and Scientific Views of Sex Hormones in 1920s America," PhD thesis, University of California Berkeley, 1997. For background on Louis Berman, I relied on Christer Nordlund, "Endocrinology and Expectations in 1930s America," *British Journal for the History of Science* 40, no. 1 (2007): 83–104.

56 **inspire four films:** Kathleen Drowne and Patrick Huber, *The 1920s* (Westport, CT: Greenwood, 2004), 25.

56 **Advice books touting endocrine cures:** "Credulity About Medicines," *Manchester Guardian*, October 8, 1925; Elizabeth Siegel Watkins, *The Estrogen*

Elixir: A History of Hormone Replacement Therapy in America (Baltimore: Johns Hopkins University Press, 2007).

56 **The pituitary was shown to release hormones:** H. Maurice Goodman, "Essays on APS Classical Papers: Discovery of Luteinizing Hormone of the Anterior Pituitary Gland," *American Journal of Physiology, Endocrinology and Metabolism* 287 (2004): E818–29.

57 **When we see misshapen:** R. G. Hoskins, "The Functions of the Endocrine Organs," *Scientific Monthly* 18, no. 3 (1924): 257–72.

58 **In the Ottoman Empire:** Richard J. Wassersug and Tucker Lieberman, "Contemporary Castration: Why the Modern Day Eunuch Remains Invisible," *British Medical Journal* 341 (2010): c4509.

58 **"Here, then . . . nervous influences":** Walter Cannon, *Bodily Changes in Pain, Hunger, Fear, and Rage* (Charleston, SC: Nabu Press, 2010), 64.

59 **"Is it possible":** Elizabeth M. Heath, "Glands as Cause of Many Crimes," *New York Times*, December 4, 1921.

59 **"Accumulating information":** Louis Berman, "Psycho-endocrinology," *Science* 67, no. 1729 (1928): 195.

60 **"My dear Rabbi Ben Ezra":** Louis Berman to Ezra Pound, "Ezra Pound Papers 1885–1976," 1925–1926, Yale Collection of American Literature, Beinecke Rare Book and Manuscript Library, YCAL MSS 43.

60 **"adrenal-centered":** Louis Berman, *The Glands Regulating Personality: A Study of Internal Secretion in Relation to the Types of Human Nature*, 2nd ed. (New York: Macmillan, 1928), 165.

60 **"will also be aggressive":** Ibid., 171.

61 **"ideal normal":** Louis Berman, *New Creations in Human Beings* (New York: Doubleday, Doran, 1938), 18.

61 **"We will be able . . . 'ideal type'":** "16-Foot Men Held a Gland Possibility," *New York Times*, December 16, 1931.

61 **in the 1920s:** Drowne and Huber, *The 1920s*, 25.

62 **Berman wasn't the only one:** Watkins, *The Estrogen Elixir*, 140; G. W. Carnrick and Co., *Organotherapy in General Practice* (Baltimore: The Lord Baltimore Press, 1924).

62 **"We are the creatures":** Chandak Sengoopta, *The Most Secret Quintessence of Life: Sex, Glands, and Hormones 1850–1950* (Chicago: University of Chicago Press, 2006), 70.

62 **"Thyroxin, parathyroid":** Louis Berman, "Crime and the Endocrine Glands," *American Journal of Psychiatry* 89, no. 2 (1932): 215–38.

63 **"should be taken with a considerable dose":** Francis Birrell, "Book Review: *The Glands Regulating Personality* by Louis Berman," *International Journal of Ethics* 32, no. 4 (1922): 450–51.

63 **"mixture of fact":** Elmer L. Severinghaus, "Review," *American Sociological Review* 4, no. 1 (1939): 144–45.

63 "For a clearer and illuminating account": Margaret Sanger, *The Pivot of Civilization* (New York: Brentano's, 1922), 236.

63 "Every truth needs men": H. L. Mencken, "Turning the Leaves with G.S.V.: A Trumpeter of Science," *American Monthly* 17, no. 6 (1925).

63 "fact mixed with fancy": Benjamin Harrow, *Glands in Health and Disease* (New York: E. P. Dutton, 1922).

63 The Second International Congress on Eugenics: Charles Benedict Davenport, "Research in Eugenics," in Charles B. Davenport et al., eds., *Scientific Papers of the Second International Congress of Eugenics*, vol. 1: *Eugenics, Genetics, and the Family* (1923): 25.

64 Dr. Sadler told colleagues: William S. Sadler, "Endocrines, Defective Germ-Plasm, and Hereditary Defectiveness," in ibid., 349.

64 *Buck v. Bell*: Buck v. Bell, 274 U.S. 200 (1927), available at: https://supreme.justia.com/cases/federal/us/274/200/case.html.

64 "We may now look forward": Berman, *The Glands Regulating Personality*, 28.

64 "Christianity is dead": Louis Berman, *The Religion Called Behaviorism* (New York: Boni and Liveright, 1927), 41.

65 three-year investigation . . . in Sing-Sing: Berman, "Crime and the Endocrine Glands"; W. H. Howell, "Crime and Disturbed Endocrine Function," *Science* 76, no. 1974 (1932): 8–9.

65 presented his findings: Berman, "Crime and the Endocrine Glands."

65 "Every criminal should be examined": Ibid., 233.

66 metabolimeter:; Frank Berry Sanborn, ed., *Basal Metabolism, Its Determination and Application* (Boston: Sanborn, 1922), 104.

67 "These are methods": Berman, "Crime and the Endocrine Glands," 10.

68 "infantile emotional characteristics": "Excerpts from the Psychiatric ('Alienist') Testimony in the Leopold Loeb Hearing," http://famous-trials.com/leopoldandloeb/1752-psychiatrictestimony.

68 Harold Hulbert: The trial proceedings can be accessed in the Clarence Darrow Digital Collection, University of Minnesota Law Library, http://moses.law.umn.edu/darrow/trials.php?tid=1.

69 Bowman–Hulbert report: Karl Bowman and Harold S. Hulbert, "Nathan Leopold Psychiatric Statement," available at http://exhibits.library.northwestern.edu/archives/exhibits/leoloeb/leopold_psych_statement.pdf and in "Loeb–Leopold Case: Psychiatrists' Report for the Defense," *Journal of Criminal Law and Criminology* 15, no. 3 (1925): 360–78. "Loeb–Leopold Murder of Franks in Chicago May 21, 1924," ibid., 347–59, gives the chronology of events.

69 "third eye": Gert-Jan Lokhorst, "Descartes and the Pineal Gland," in *The Stanford Encyclopedia of Philosophy* (2015), https://plato.stanford.edu/entries/pineal-gland/; Mark S. Morrisson, "'Their Pineal Glands Aglow': Theosophical Physiology in 'Ulysses'," *James Joyce Quarterly* 46, no. 3–4 (2008), 509–27.

70 **"remove the ordinary restraint":** Edward Tenner, "The Original Natural Born Killers," *Nautilus*, September 11, 2014.

70 **"the so-called trial":** Higdon, *Leopold and Loeb*, 164.

70 **"in their applicability":** Judge Caverly's decision and sentence available at: http://famous-trials.com/leopoldandloeb/1747-judgedecision.

5. THE VIRILE VASECTOMY

I have drawn from Nordlund, "Endocrinology and Expectations in 1930s America," Rechter, "The Glands of Destiny," and Sengoopta, *The Most Secret Quintessence of Life*, for background information on hormone research in the 1920s and 1930s. Details on Steinach are drawn from Eugen Steinach, *Sex and Life: Forty Years of Biological Experiments* (New York: Viking, 1940), and Chandak Sengoopta, "Tales from the Vienna Labs: The Eugen Steinach–Harry Benjamin Correspondence," *Newsletter of the Friends of the Rare Book Room, New York Academy of Medicine*, no. 2 (Spring, 2000): 1–2, 5–9. More information about John Brinkley can be found in R. Alton Lee, *The Bizarre Careers of John R. Brinkley* (Lexington: University Press of Kentucky, 2002), and Pope Brock, *Charlatan: America's Most Dangerous Huckster, the Man Who Pursued Him, and the Age of Flimflam* (New York: Broadway Books, 2009). For more on Charles Édouard Brown–Séquard, see Aminoff, *Brown-Séquard*.

73 **without Steinach in the operating room:** Michael A. Kozminski and David A. Bloom, "A Brief History of Rejuvenation Operations," *Journal of Urology* 187, no. 3 (2012): 1130–34.

73 **questionable procedures:** "Paris Scientist Tells of Gland Experiments," *Los Angeles Times*, June 5, 1923; "New Ponce De Leon Coming," *Baltimore Sun*, September 16, 1923; "Gland Treatment Spreads in America," *New York Times*, April 8, 1923.

74 **"Much of it the result":** Hans Lisser to Dr. Cushing, July 19, 1921, in Yale University Medical School archives, HC Reprints X, no. 156.

74 **considered a proper scientist:** See https://www.nobelprize.org/nomination/archive/show_people.php?id=8765.

75 **theories that had been bandied about for centuries:** Kozminski and Bloom, "A Brief History of Rejuvenation Operations." The rationale for Steinach's procedure is also provided in E. Steinach, "Biological Methods Against the Process of Old Age," *Medical Journal and Record* 125, no. 2345 (1927): 78–81, 161–64.

75 **"fantastic experiments":** Steinach, *Sex and Life*, 49.

75 **"must be regarded":** Ibid., 49–50.

75 **the treatment du jour:** "Elixir of Life: The Brown-Sequard Discovery," *Aroha and Ohinemu News and Upper Thames Advocate*, September 25, 1889.

76 **"revived my creative power":** Chandak Sengoopta, "Glandular Politics: Experimental Biology, Clinical Medicine, and Homosexual Emancipation in Fin-de-Siècle Central Europe," *Isis* 89, no. 3 (1998): 445–73; Chandak

Sengoopta, "'Dr Steinach coming to make old young!': Sex Glands, Vasectomy and the Quest for Rejuvenation in the Roaring Twenties," *Endeavour* 27, no. 3 (2003): 122–26.

76 **"my memory is better":** Steinach and Loebel, *Sex and Life*, 173.

76 **Self-help books sold wildly:** Drowne and Huber, *The 1920s*; Michael Pettit, "Becoming Glandular: Endocrinology, Mass Culture, and Experimental Lives in the Interwar Age," *American Historical Review* 118, no. 4 (2013): 1052–76.

77 **"technology of the self":** Pettit, "Becoming Glandular," 5.

77 **Steinach had not set out to devise a blockbuster:** Laura Davidow Hirschbein, "The Glandular Solution: Sex, Masculinity, and Aging in the 1920s," *Journal of the History of Sexuality* 9, no. 3 (2000): 277–304.

77 **study of frog sex:** Sengoopta, *The Most Secret Quintessence of Life*, 57.

78 **"But it seemed to me," "what I actually saw":** Steinach and Loebel, *Sex and Life*, 16.

79 **"the whole complicated phenomenon":** Ibid., 3.

79 **"Everyone knows":** Ibid., 39.

80 **testicles work no matter where they dangle:** Per Södersten et al., "Eugen Steinach: The First Neuroendocrinologist," *Endocrinology* 155, no. 3 (2014): 688–95.

80 **"All the male rats":** Steinach and Loebel, *Sex and Life*, 30.

81 **"Without hesitation":** Ibid., 32.

82 **"with the same care . . . definitely male":** Ibid., 64.

82 **"erotization":** Steinach titles a chapter of *Sex and Life* "Experiments in Explanation and Erotization," and writes, "I have coined the expression 'erotization of the central nervous system' or 'erotization'" (30).

82 **"The most important decision":** Ibid., 71.

83 **Karl Kraus:** Christopher Turner, "Vasectomania, and Other Cures for Sloth," *Cabinet*, no. 29: Spring 2008.

83 **pseudo-homosexual behavior:** Sengoopta, *The Most Secret Quintessence of Life*, 80.

83 **the adjacent tissue would overcompensate:** Kozminski and Bloom, "A Brief History of Rejuvenation Operations."

84 **His patient was Anton W.:** Stephen Lock, "'O That I Were Young Again': Yeats and the Steinach Operation," *British Medical Journal* (Clinical Research Edition) 287, no. 6409 (1983): 1964–68.

84 **"extraordinary improvement":** Steinach and Loebel, *Sex and Life*, 178.

85 **Journalists loved the story:** "Gland Treatment Spreads in America," *New York Times*, April 8, 1923; "New Ponce De Leon Coming," *Baltimore Sun*, September 16, 1923.

85 **"We have been Voronoffed":** Van Buren Thorne, "The Craze for Rejuvenation," *New York Times*, June 4, 1922.

85 **"hocus pocus":** Morris Fishbein, *Fads and Quackery in Healing: An Analysis of the Foibles of the Healing Cults, With Essays on Various Other Peculiar Notions in the Health Field* (New York: Covici, Friede, 1932).

87 "How I was Made Twenty Years Younger": Angus McLaren, *Reproduction by Design* (Chicago: University of Chicago Press, 2012), 85–86; Van Buren Thorne, "Dr. Steinach and Rejuvenation," *New York Times*, June 26, 1921.

87 "some disharmony": McLaren, *Reproduction by Design*, 86.

88 created the market for hormone-based rejuvenation: Södersten et al., "Eugen Steinach."

88 plenty of serious science: E. C. Hamblen, "Clinical Experience with Follicular and Hypophyseal Hormones," *Endocrinology* 15, no. 3 (1931): 184–94; Michael J. O'Dowd and Elliot E. Phillips, "Hormones and the Menstrual Cycle," *The History of Obstetrics and Gynaecology* (New York: Pantheon, 1994), 255–75.

6. SOUL MATES IN SEX HORMONES

This chapter is based on extensive interviews with Dr. Howard W. Jones, Jr., his children, and his colleagues, including the Joneses' longtime assistant Nancy Garcia; Mary F. Davies, president of the Jones Foundation; Dr. Edward Wallach, professor emeritus of gynecology and obstetrics, Johns Hopkins University School of Medicine; Dr. Alan DeCherney, senior investigator in reproductive endocrinology and science, National Institutes of Health; Dr. Claude Migeon, pediatric endocrinologist, Johns Hopkins University School of Medicine; and Dr. Robert Blizzard, professor emeritus of pediatric endocrinology, University of Virginia. I read through Dr. Jones's personal archive, which includes photographs, correspondence, publications, and unpublished memoirs, and consulted the papers of Arthur Hertig, the Harvard professor of pathology whose lab provided the placenta that led to Georgeanna Jones's discovery (kindly made available by his son Andrew Hertig).

90 "I, of course, thought": Howard W. Jones, Jr., life story, Jones archive.

92 recently published medical book: Edgar Allen, ed., *Sex and Internal Secretions: A Survey of Recent Research* (Baltimore: Williams and Wilkins, 1932).

94 the A–Z test was the pregnancy test: Henry W. Louria and Maxwell Rosenzweig, "The Aschheim–Zondek Hormone Test for Pregnancy," *Journal of the American Medical Association* 91, no. 25 (1928): 1988; "Aschheim and Zondek's Test for Pregnancy," *British Medical Journal* (1929): 232C; "The Zondek–Ascheim Test for Pregnancy," *Canadian Medical Association Journal* 22, no. 2 (1930): 251–53; George H. Morrison, "Zondek and Aschheim Test for Pregnancy," *Lancet* 215, no. 5551 (1930): 161–62.

95 Earl Engle: Howard W. Jones, Jr., "Chorionic Gonadotropin: A Narrative of Its Identification and Origin and the Role of Georgeanna Seegar Jones," *Obstetrical and Gynecological Survey* 62, no. 1 (2007): 1–3.

95 in the placenta as well: Ibid.

96 A do-it-yourself kind of person: Michael Rogers, "The Double-Edged Helix,"

Rolling Stone, March 25, 1976; Rebecca Skloot, *The Immortal Life of Henrietta Lacks* (New York: Crown Publishers, 2010); Jane Maienschein, Marie Glitz, Garland E. Allen, eds., *Centennial History of the Carnegie Institution of Washington*, vol. 5 (Cambridge, UK: Cambridge University Press, 2005), 143.

96 **The movement pushed the cells:** Andrew Artenstein, ed., *Vaccines: A Biography* (New York: Springer, 2010), 152.

96 **Pulses of carbon dioxide:** Duncan Wilson, *Tissue Culture in Science and Society: The Public Life of a Biological Technique in Twentieth-Century Britain* (London: Palgrave Macmillan, 2011), 60.

98 **not the pituitary:** Jones, "Chorionic Gonadotropin."

98 **The letter was published:** George Gey, G. Emory Seegar, and Louis M. Hellman, "The Production of a Gonadotrophic Substance (Prolan) by Placental Cells in Tissue Culture," *Science* 88, no. 2283 (1938): 306–7. For a history of the experiment, see Jones, "Chorionic Gonadotropin."

99 **"Georgeanna is the most important":** Dr. Howard W. Jones, Jr., author interview.

100 **One patient remembered her:** Frances Neal to Howard W. Jones, Jr., condolence card, 2005, Jones archive.

7. MAKING GENDER

This chapter is based on extensive interviews with Bo Laurent, who shared her medical records with me, and with Dr. Arlene Baratz, physician and medical advisor to the Androgen Insensitivity Support Group; Dr. Katie Baratz, psychiatrist; Georgiann Davis, assistant professor of sociology, University of Nevada; and several other people who talked about how intersexuality affected their lives, as well as with endocrinologists who cared for intersex patients both in the 1950s and today. I had access to medical records (with names redacted) from the care of intersex children at Columbia University in the 1930s and 1940s, to the papers of John Money at the Kinsey Institute, and to notes from meetings about intersex children in the personal archives of Dr. Howard W. Jones, Jr. I interviewed experts including Dr. Claude Migeon and Dr. Howard W. Jones, Jr., of Johns Hopkins and David Sandberg, PhD, clinical psychologist, University of Michigan; and the historians Dr. Sandra Eder, assistant professor, University of California, Berkeley, Dr. Elizabeth Reis, professor, Macaulay Honors College, City University of New York, and Dr. Katrina Karkazis, senior research scholar, Center for Biomedical Ethics, Stanford University. Further background information comes from: Alice Dreger, *Hermaphrodites and the Medical Invention of Sex* (Cambridge, MA: Harvard University Press: 1998); Alice Dreger, *Intersex in the Age of Ethics* (Hagerstown, MD: University Publishing Group, 1999); Katrina Karkazis, *Fixing Sex: Intersex, Medical Authority, and Lived Experience* (Durham, NC: Duke University Press, 2008); Elizabeth Reis, *Bodies in Doubt: An American History of Intersex* (Baltimore: Johns Hopkins University Press, 2009); Sandra

Eder, "The Birth of Gender: Clinical Encounters with Hermaphroditic Children at Johns Hopkins (1940–1956)," PhD thesis in the history of medicine, Johns Hopkins University, 2011; Suzanne J. Kessler, *Lessons from the Intersexed* (New Brunswick, NJ: Rutgers University Press: 2002); Georgiann Davis, *Contesting Intersex: The Dubious Diagnosis* (New York: New York University Press, 2015); Hida Viloria, *Born Both: An Intersex Life* (New York: Hachette, 2017); Thea Hillman, *Intersex (for lack of a better word)* (San Francisco: Manic D Press, 2008); and Cheryl Chase, "Hermaphrodites with Attitude: Mapping the Emergence of Intersex Political Activism," *GLQ: A Journal of Lesbian and Gay Studies* 4, no. 2 (1998): 189–211.

108 **"The last decade has witnessed":** Howard W. Jones, Jr., and Lawson Wilkins, "Gynecological Operations in 94 patients with Intersexuality: Implications Concerning the Endocrine Theory of Sexual Differentiation," *American Journal of Obstetrics and Gynecology* 82, no. 5 (1961): 1142–53.

109 **Hermaphroditus:** Howard W. Jones, Jr., and William Wallace Scott, *Hermaphroditism, Genital Anomalies and Related Endocrine Disorders* (Baltimore: Williams and Wilkins, 1958); Anne Fausto-Sterling, "The Five Sexes," *Sciences* 33, no. 2 (1993): 20–24.

110 **Today, ambiguous genitalia:** M. Blackless et al., "How Sexually Dimorphic Are We? Review and Synthesis," *American Journal of Human Biology* 12, no. 2 (2000): 151–66; Gerald Callahan, *Between XX and XY: Intersexuality and the Myth of Two Sexes* (Chicago: Chicago Review Press, 2009); Diane K. Wherrett, "Approach to the Infant with a Suspected Disorder of Sex Development," *Pediatric Clinics of North America* 62, no. 4 (2015): 983–99.

110 **"Every zygote":** Edgar Allen, ed., *Sex and Internal Secretions: A Survey of Recent Research* (Baltimore: Williams and Wilkins, 1932), 5.

111 **anti-Mullerian hormone:** N. Josso, "Professor Alfred Jost: The Builder of Modern Sex Differentiation," *Sexual Development* 2, no. 2 (2008): 55–63.

111 **femaleness may not be merely a default:** Rebecca Jordan-Young, *Brain Storm: The Flaws in the Science of Sex Difference* (Cambridge, MA: Harvard University Press, 2010), 25.

111 **females are created by a passive process:** H. H. Yao, "The Pathway to Femaleness: Current Knowledge on Embryonic Development of the Ovary," *Molecular and Cellular Endocrinology* 230, no. 1–2 (2005): 87–93.

112 **a study showing that cortisone helped children:** Howard W. Jones, Jr., and Georgeanna E. S. Jones, "The Gynecological Aspects of Adrenal Hyperplasia and Allied Disorders," *American Journal of Obstetrics and Gynecology* 68, no. 5 (1954): 1330–65.

113 **"therapeutic tour de force":** Paul Gyorgy et al., "Inter-University Round Table Conference by the Medical Faculties of the University of Pennsylvania and Johns Hopkins University: Psychological Aspects of the Sexual

Orientation of the Child with Particular Reference to the Problem of Intersexuality," *Journal of Pediatrics* 47, no. 6 (1955): 771–90.

113 **John Money:** Secondary sources include Terry Goldie, *The Man Who Invented Gender: Engaging Ideas of John Money* (Vancouver: UBC Press, 2014); Karkazis, *Fixing Sex*; and John Money, "Intersexual Problems," in Kenneth Ryan and Robert Kistner, eds., *Clinical Obstetrics and Gynecology* (Baltimore: Harper & Row, 1973).

113 **"fuckology":** Iain Morland, "Pervert or Sexual Libertarian? Meet John Money, 'the father of f*ology,' " *Salon*, January 4, 2014; also see Lisa Downing, Iain Morland, and Nikki Sullivan, *Fuckology* (Chicago: Chicago University Press: 2015).

113 **"There have been many illustrious":** Richard Green and John Money, "Effeminacy in Prepubertal Boys," *Pediatrics* 27, no. 286 (1961): 286–91.

114 **widely publicized court case:** Testimony of Dr. John William Money in Joseph Acanfora III v. Board of Education of Montgomery County, Montgomery County Public Schools, U.S. District Court for the District of Maryland – 359 F. Supp. 843 (1973).

114 **panel about sexuality sponsored by *Playboy*:** "New Sexual Lifestyles: A symposium on emerging behavior patterns from open marriage to group sex," *Playboy*, September 1973.

114 **seven criteria:** Howard W. Jones, Jr., "Hermaphroditism," *Progress in Gynecology* 3 (1957): 35–49; Lawson Wilkins et al., "Masculinization of the Female Fetus Associated with Administration of Oral and Intramuscular Progestins During Gestation: Non-Adrenal Female Pseudohermaphrodism," *Journal of Clinical Endocrinology and Metabolism* 18, no. 6 (1958): 559–85.

115 **"By gender role":** John Money et al., "An Examination of Some Basic Sexual Concepts: The Evidence of Human Hermaphroditism," *Bulletin of the Johns Hopkins Hospital* 97, no. 4 (1955): 301–19.

116 **importance of how a child is raised:** Karkazis, *Fixing Sex*.

116 **"There doesn't seem to be any doubt":** Dr. Joan Hampson, minutes from an American Urological Association meeting, 1956, Jones archive.

117 **condemn the practice:** Associated Press, "Pressure Mounts to Curtail Surgery on Intersex Children," *New York Times*, July 25, 2017.

117 **"bold articles were unusual":** Reis, *Bodies in Doubt*, 177.

117 **"I thought he was smart":** Dr. Milton Diamond, author interview.

117 **scathing scientific article:** Milton Diamond and H. Keith Sigmundson, "Sex Reassignment at Birth: A Long Term Review and Clinical Implications," *Archives of Pediatrics and Adolescent Medicine* 151, no. 3 (1997): 298–304.

117 **an exposé:** John Colapinto, "The true story of John/Joan," *Rolling Stone* 775 (1997): 54-73, 97; John Colapinto, *As Nature Made Him: The Boy who Was Raised as a Girl* (New York: Harper Perennial, 2000).

118 **"a nuanced analysis":** Karkazis, *Fixing Sex*, 47.

120 **read up on sexuality and gender anatomy:** C. H. Phoenix et al., "Organizing Action of Prenatally Administered Testosterone Propionate on the Tis-

sues Mediating Mating Behavior in the Female Guinea Pig," *Endocrinology* 65, no. 3 (1959): 369–82.

120 **DES:** Randi Hutter Epstein, *Get Me Out: A History of Childbirth from the Garden of Eden to the Sperm Bank* (New York: W. W. Norton, 2010).

121 **In 1993, Anne Fausto-Sterling:** Fausto-Sterling, "The Five Sexes."

121 **eradicate the label "hermaphrodite":** J. M. Morris, "Intersexuality," *Journal of the American Medical Association* 163, no. 7 (1957): 538–42; Robert B. Edgerton, "Pokot Intersexuality: An East African Example of the Resolution of Sexual Incongruity," *American Anthropologist* 66, no. 6 (1964): 1288–99; John Money, "Psychologic Evaluation of the Child with Intersex Problems," *Pediatrics* 36, no. 1 (1965): 51–55; Cheryl Chase, "Letters from Readers," *The Sciences* 33, no. 3 (1993).

123 **doctors are encouraged to speak openly:** Jennifer E. Dayner et al., "Medical Treatment of Intersex: Parental Perspectives," *Journal of Urology* 172, no. 4 (2004): 1762–65.

123 **in 2013 Swiss and German researchers:** Jürg C. Streuli et al., "Shaping Parents: Impact of Contrasting Professional Counseling on Parents' Decision Making for Children with Disorders of Sex Development," *Journal of Sexual Medicine* 10, no. 8 (2013): 1953–60.

124 **"It's true that people":** Bo Laurent, author interview.

8. GROWING UP

This chapter was based on extensive interviews with Dr. Al and Barbara Balaban, along with newspaper clippings which they generously shared with me, and interviews with Dr. Robert Blizzard, professor emeritus of pediatric endocrinology, University of Virginia; Dr. Albert Parlow, professor of hormone biochemistry, LA BioMed; Dr. Michael Aminoff, director of the Parkinson's Disease and Movement Disorders Clinic, University of California San Francisco; and Carol Hintz, the widow of Dr. Raymond Hintz. A thorough overview of the history of growth hormone treatment can be found in Stephen Hall, *Size Matters: How Height Affects the Health, Happiness, and Success of Boys—and the Men They Become* (New York: Houghton Mifflin Harcourt, 2006), Susan Cohen and Christine Cosgrove, *Normal at Any Cost: Tall Girls, Short Boys, and the Medical Industry's Quest to Manipulate Height* (New York: Jeremy P. Tarcher/Penguin, 2009), and Aimee Medeiros, *Heightened Expectations* (Tuscaloosa: University of Alabama Press, 2016), based on her PhD thesis in the history of health sciences, University of California San Francisco, 2012, which I consulted. Aurelia Minutia and Jennifer Yee shared information about Dr. Edna Sobel.

128 **Anthropologists have theorized:** Ron G. Rosenfeld, "Endocrine Control of Growth," in Noël Cameron and Barry Bogin, eds., *Human Growth and Development*, 2nd ed. (New York: Elsevier, 2012).

129 **a hormone could "cure" shortness:** Melvin Grumbach, "Herbert McLean

Evans, Revolutionary in Modern Endocrinology: A Tale of Great Expectations," *Journal of Clinical Endocrinology and Metabolism* 55, no. 6 (1982): 1240–47.

129 **Dr. Oscar Riddle:** "Scientist Predicts Pituitary Treatment Will Overcome the 'Inferiority Complex,'" *New York Times*, August 2, 1937.

129 **"life of hellish dwarfism":** Medeiros, "Heightened Expectations" (PhD thesis), 152.

129 **immaturity and insecurity:** Sheila Rothman and David Rothman, *The Pursuit of Perfection: The Promise and Perils of Medical Enhancement* (New York: Pantheon, 2003), 173.

129 **"The combination of endocrinology":** Ibid., 174.

130 **stories about growth hormone breakthroughs:** "Hormone to Aid Growth Isolated, But It Is Too Costly for Wide Use," *New York Times*, March 8, 1944; "What Scientists Are Doing," *New York Herald Tribune*, March 19, 1944; Choh Hao Li and Herbert Evans, "The Isolation of Pituitary Growth Hormone," *Science* 99, no. 2566 (1944): 183–84.

130 **In 1958, newspapers wrote about a cure:** Earl Ubell, "Hormone Makes Dwarf Grow: May Also Offer Clues in Cancer, Obesity, Aging," *New York Herald Tribune*, March 29, 1958; Earl Ubell, "Hormones Now May Be Tailor-Made," *New York Herald Tribune*, May 10, 1959.

132 **testosterone didn't increase growth:** Edna Sobel et al., "The Use of Methyl-testosterone to Stimulate Growth: Relative Influence on Skeletal Maturation and Linear Growth," *Journal of Clinical Endocrinology and Metabolism* 16, no. 2 (1956): 241–48.

136 **The Evans–Li study:** Li and Evans, "The Isolation of Pituitary Growth Hormone."

137 **Dr. Maurice Raben:** M. S. Raben, "Letters to the Editor: Treatment of a Pituitary Dwarf with Human Growth Hormone," *Journal of Clinical Endocrinology and Metabolism* 18, no. 8 (1958): 901–3.

137 **"Hormone Makes Dwarf Grow":** Earl Ubell, "Hormone Makes Dwarf Grow," *New York Herald Tribune*, March 29, 1958.

137 **"won't produce basketball players":** Alton L. Blakeslee, "Stimulant Found in Pituitary Powder: Growth Hormone Isolated: Found Capable of Inducing Added Height in Children Dwarfed by Natural Causes," *Pittsburgh Post-Gazette*, March 29, 1958.

140 **half-gallon milk container:** Dr. Salvatore Raiti, author interview.

145 **publicize their cause:** Rothman and Rothman, *The Pursuit of Perfection*, 171.

145 **"otherwise, it would have been jungle warfare":** Podine Schoenberger, "Pilot Honored by Pathologists," *New Orleans Times-Picayune*, March 26, 1968.

145 **The agency also issued guidelines:** Ibid.

145 **natural, safer choice:** Robert Blizzard, "History of Growth Hormone Therapy," *Indian Journal of Pediatrics* 79, no. 1 (2012): 87–91.

146 **"We Can End Dwarfism":** Medeiros, "Heightened Expectations" (PhD thesis), 166.

9. MEASURING THE IMMEASURABLE

Dr. Thomas Foley, professor of pediatric endocrinology, University of Pittsburgh, provided background on thyroid history. Details of Rosalyn Yalow's life are drawn from *Rosalyn Yalow, Nobel Laureate: Her Life and Work in Medicine* (New York: Basic Books, 1998) by a former student turned colleague and family friend, Dr. Eugene Straus. I also interviewed several of Dr. Yalow's colleagues, as well as her children, and viewed home video clips of Yalow, events in her honor, and memorial events.

150 **"bear down"**: Straus, *Rosalyn Yalow*, 46.

150 **"They had to have a war"**: Ibid., 34.

151 **"She pushed me"**: Mildred Dresselhaus, home video of a memorial service, Yalow archive.

152 janitor'sclosetintoalaboratory:"RosalynYalowandSolomonBerson,"Chemical Heritage Foundation, August 13, 2015, https://www.sciencehistory .org/historical-profile/rosalyn-yalow-and-solomon-a-berson.

154 **The article was published in 1956**: S. A. Berson and R. S. Yalow et al., "Insulin-I^{131} Metabolism in Human Subjects: Demonstration of Insulin Binding Globulin in the Circulation of Insulin-Treated Subjects," *Journal of Clinical Investigation* 35 (1956): 170–90.

156 **a 1960 article**: Rosalyn S. Yalow and Solomon A. Berson, "Immunoassay of Endogenous Plasma Insulin in Man," *Journal of Clinical Investigation* 39, no. 7 (1960): 1157–75.

158 **"Fortunately, that is not difficult"**: Ruth H. Howes, "Rosalyn Sussman Yalow (1921–2011)," American Physical Society Sites: Forum on Physics and Society, 2015.

158 **"Initially . . . new ideas are rejected"**: Endocrine Society Staff, "In Memoriam: Dr. Rosalyn Yalow, PhD, 1921–2011," *Molecular Endocrinology* 26, no. 5 (2012): 713–14.

158 **She died on May 30, 2011**: Denise Gellene, "Rosalyn S. Yalow, Nobel Medical Physicist, Dies at 89," *New York Times*, June 1, 2011.

10. GROWING PAINS

Background details are drawn from Jennifer Cooke, *Cannibals, Cows and the CJD Catastrophe* (Sydney: Random House Australia, 1998). I also relied on Susan Cohen and Christine Cosgrove, *Normal at Any Cost: Tall Girls, Short Boys, and the Medical Industry's Quest to Manipulate Height* (New York: Jeremy P. Tarcher/ Penguin, 2009). This book covers growth hormone, and also provides a history of giving estrogen to stunt the growth of girls considered too tall. I conducted many interviews with growth hormone patients, FDA officials, and doctors familiar with the tragedy and the biology of CJD, including Carol Hintz (the widow of Dr. Raymond Hintz); Dr. Michael Aminoff; Dr. Robert Blizzard; Dr. Albert

Parlow; Dr. Robert Rohwer, associate professor of neurology, University of Maryland; Dr. Paul Brown, senior investigator, National Institutes of Health; Dr. Alan Dickinson, founder of the neuropathogen unit, University of Edinburgh; and Dr. Judith Fradkin, director of the division of diabetes, endocrinology, and metabolic diseases, National Institutes of Health. The journalist Emily Green generously shared not only her coverage of the growth hormone–CJD story in the U.K. but her sources as well. Nicholas Smith, a former student of mine, translated the French newspapers into English for me.

160 **Joey Rodriguez:** Thomas Koch et al., "Creutzfeldt-Jakob Disease in a Young Adult with Idiopathic Hypopituitarism: Possible Relation to the Administration of Cadaveric Human Growth Hormone," *New England Journal of Medicine* 313, no. 12 (1985): 731–33.

161 **"didn't need to go for a spin":** Cooke, *Cannibals, Cows and the CJD Catastrophe*, 110.

167 **"The effect of the new information":** Paul Brown, "Reflections on a Half-Century in the Field of Transmissible Spongiform Encephalopathy," *Folia Neuropathologica* 47, no. 2 (2009): 95–103.

167 **"Only Genentech is not in mourning":** Paul Brown et al., "Potential Epidemic of Creutzfeldt-Jakob Disease from Human Growth Hormone Therapy," *New England Journal of Medicine* 313, no. 12 (1985): 728–31; Paul Brown, "Human Growth Hormone Therapy and Creutzfeldt-Jakob Disease: A Drama in Three Acts," *Pediatrics* 81 (1988): 85–92; Paul Brown, "Iatrogenic Creutzfeldt-Jakob Disease," *Neurology* 67, no. 3 (2006): 389–93.

170 **"Once a year—tops":** David Davis, "Growing Pains," *LA Weekly*, March 21, 1997.

171 **appeared to confirm Parlow's fears:** Joseph Y. Abrams et al., "Lower Risk of Creutzfeldt-Jakob Disease in Pituitary Growth Hormone Recipients Initiating Treatment after 1977," *Journal of Clinical Endocrinology and Metabolism* 96, no. 10 (2011): E1666–69; Genevra Pittman, "Purified Growth Hormone Not Tied to Brain Disease," Reuters Health, August 19, 2011.

172 **33 confirmed deaths:** Dr. Larry Schonberger, Centers for Disease Control, email to author, October 24, 2017, and Christine Pearson, CDC spokesperson, email to author, October 5, 2017. The 33 deaths include one case related to hormone made by a pharmaceutical firm. Other potential cases have been reported, including the 2013 death of a child denied treatment by the U.S. government program because he didn't meet the height criteria who was given hormones from Europe, reported in Brian S. Appleby et al., "Iatrogenic Creutzfeldt-Jakob Disease from Commercial Cadaveric Human Growth Hormone," *Emerging Infectious Diseases* 19, no. 4 (2013): 682–84.

172 **In the U.K., 78 deaths:** Dr. Peter Rudge, email to author, October 4, 2017.

See also P. Rudge et al., "Iatrogenic CJD Due to Pituitary-Derived Growth Hormone with Genetically Determined Incubation Times of Up to 40 Years," *Brain* 138, no. 11 (2015): 3386–99.

172 **British courts ruled:** Emily Green, "A Wonder Drug That Carried the Seeds of Death," *Los Angeles Times*, May 21, 2000.

172 **a group of French families sued:** Several articles have been written about the French lawsuits. See Angelique Chrisafis, "French Doctors on Trial for CJD Deaths after Hormone 'Misuse,'" *Guardian*, February 6, 2008; Barbara Casassus, "INSERM Doubts Criminality in Growth Hormone Case," *Science* 307, no. 5716 (2005): 1711, and "Acquittals in CJD Trial Divide French Scientists," *Science* 323, no. 5913 (2009): 446; Pierre-Antoine Souchard and Verena Von Derschau, "6 Acquitted in French Trial over Hormone Deaths," Associated Press, in *San Diego Union-Tribune*, January 14, 2009.

11. HOTHEADS: THE MYSTERIES OF MENOPAUSE

Mary Jane Minkin, clinical professor of obstetrics, gynecology, and reproductive services, Yale University, provided professional expertise on the subject of menopause. I also interviewed numerous researchers and clinicians, including Dr. Lila Nachtigall, professor of obstetrics and gynecology, New York University; Dr. Hugh Taylor, chief of obstetrics and gynecology, Yale University; Dr. Nanette Santoro, professor of obstetrics and gynecology, University of Colorado School of Medicine; and Cindy Pearson, executive director of the Women's Health Network. Several menopausal women were willing to speak openly about their symptoms, among them one woman—just one—who said she'd never felt better than when menopause hit.

174 **coauthored *Menopause*:** Charles B. Hammond et al., *Menopause: Evaluation, Treatment, and Health Concerns—Proceedings of a National Institutes of Health Symposium Held in Bethesda, Maryland, April 21–22, 1988* (New York: Alan R. Liss, 1989).

176 **"And the biological changes":** Helen E. Fisher, "Mighty Menopause," *New York Times*, October 21, 1992.

177 **symptoms linger for decades:** F. Kronenberg, "Menopausal Hot Flashes: A Review of Physiology and Biosociocultural Perspective on Methods of Assessment," *Journal of Nutrition* 140, no. 7 (2010): 1380s–85s.

177 **slipped into a few sitcoms:** Elizabeth Siegel Watkins, *The Estrogen Elixir: A History of Hormone Replacement Therapy in America* (Baltimore: Johns Hopkins University Press, 2007).

177 **A few NIH studies:** Ibid., 244; Nancy Krieger et al., "Hormone Replacement Therapy, Cancer, Controversies, and Women's Health: Historical, Epidemiological, Biological, Clinical, and Advocacy Perspectives," *Journal*

of Epidemiology and Community Health 59, no. 9 (2005): 740–48; Watkins, *The Estrogen Elixir*, 244; A. Heyman et al., "Alzheimer's Disease: A Study of Epidemiological Aspects," *Annals of Neurology* 15, no. 4 (1984): 335–41; M. X. Tang et al., "Effect of Oestrogen During Menopause on Risk and Age at Onset of Alzheimer's Disease," *Lancet* 348, no. 9025 (1996): 429–32.

178 **Clues were beginning to emerge:** Margaret Morganroth Gullette, "What, Menopause Again?" *Ms.*, July 1993, 34; Nancy Fugate Woods, "Menopause: Models, Medicine, and Midlife," *Frontiers* 19, no. 1 (1998): 5–19.

178 **Dr. Robert Freedman:** Dr. Robert Freedman, author interview; Robert R. Freedman, "Biochemical, Metabolic, and Vascular Mechanisms in Menopausal Hot Flashes," *Fertility and Sterility* 70, no. 2 (1998): 332–37, and "Menopausal Hot Flashes: Mechanisms, Endocrinology, Treatment," *Journal of Steroid Biochemistry and Molecular Biology* 142 (2014): 115–20. See also Denise Grady, "Hot Flashes: Exploring the Mystery of Women's Thermal Chaos," *New York Times*, September 3, 2002.

180 **not yet known how they are connected:** Kronenberg, "Menopausal Hot Flashes."

180 **killer whales have hot flashes:** Lauren Brent, author interview; Lauren Brent et al., "Ecological Knowledge, Leadership, and the Evolution of Menopause in Killer Whales," editorial comment, *Obstetrical and Gynecological Survey* 70, no. 11 (2015): 701–2.

182 **she collected three brains:** Naomi Rance, author interview; Naomi E. Rance et al., "Modulation of Body Temperature and LH Secretion by Hypothalamic KNDy (kisspeptin, neurokinin B and dynorphin) Neurons: A Novel Hypothesis on the Mechanism of Hot Flushes," *Frontiers in Neuroendocrinology* 34, no. 3 (2013): 211–27; N. E. Rance et al., "Postmenopausal Hypertrophy of Neurons Expressing the Estrogen Receptor Gene in the Human Hypothalamus," *Journal of Clinical Endocrinology and Metabolism* 71, no. 1 (1990): 79–85.

182 **she studied six more brains:** N. E. Rance and W. S. Young III, "Hypertrophy and Increased Gene Expression of Neurons Containing Neurokinin-B and Substance-P Messenger Ribonucleic Acids in the Hypothalami of Postmenopausal Women," *Endocrinology* 128, no. 5 (1991): 2239–47. For a review of the Rance research, see Ty William Abel and Naomi Ellen Rance, "Stereologic Study of the Hypothalamic Infundibular Nucleus in Young and Older Women," *Journal of Comparative Neurology* 424, no. 4 (2000): 679–88.

182 **injections of neurokinin-B:** Channa Jayasena et al., "Neurokinin B Administration Induces Hot Flushes in Women," *Scientific Reports* 5, no. 8466 (2015).

183 **a drug that blocks neurokinin-B:** Julia K. Prague et al., "Neurokinin 3 Receptor Antagonism as a Novel Treatment for Menopausal Hot Flushes: A Phase 2, Randomised, Double-Blind, Placebo-Controlled Trial," *Lancet* 389, no. 10081 (May 2017): 1809–20. Articles on the potential new non-

hormone drug include Megan Cully, "Neurokinin 3 Receptor Antagonist Revival Heats Up with Astellas Acquisition," *Nature Reviews Drug Discovery* 16, no. 6 (2017): 377.

183 **another group of hormone researchers:** Heyman et al., "Alzheimer's Disease"; V. W. Henderson et al., "Estrogen Replacement Therapy in Older Women: Comparisons Between Alzheimer's Disease Cases and Nondemented Control Subjects," *Archives of Neurology* 51, no. 9 (1994): 896–900; Tang et al., "Effect of Oestrogen."

183 **white and upper-class:** Randall S. Stafford et al., "The Declining Impact of Race and Insurance Status on Hormone Replacement Therapy," *Menopause* 5, no. 3 (1998): 140–44; Watkins, *The Estrogen Elixir.*

183 **black women were 60 percent less likely:** Kate M. Brett and Jennifer H. Madans, "Differences in Use of Postmenopausal Hormone Replacement Therapy by Black and White Women," *Menopause* 4, no. 2 (1997): 66–76.

184 **data from more than 30,000 office visits:** Stafford et al., "The Declining Impact of Race and Insurance Status."

184 **a two-day conference in 2004:** Krieger et al., "Hormone Replacement Therapy, Cancer, Controversies, and Women's Health."

184 **"No woman can escape":** Robert Wilson, *Feminine Forever* (New York: Pocket Books, 1968), 52.

184 **funded by three drug companies:** Krieger et al., "Hormone Replacement Therapy, Cancer, Controversies, and Women's Health"; Judith Houck, *Hot and Bothered: Women, Medicine, and Menopause in Modern America* (Cambridge, MA: Harvard University Press, 2006).

186 **Prescriptions . . . nearly halved:** Krieger et al., "Hormone Replacement Therapy, Cancer, Controversies, and Women's Health."

187 **PEPI:** The Writing Group for the PEPI Trial, "Effects of estrogen or estrogen/progestin regimens on heart disease risk factors in postmenopausal women: The Postmenopausal Estrogen/Progestin Interventions (PEPI) Trial," *Journal of the American Medical Association* 273, no. 3 (1995): 199–208.

187 **Another massive study:** Meir J. Stampfer et al., "Postmenopausal Estrogen Therapy and Cardiovascular Disease," *New England Journal of Medicine* 325, no. 11 (1991): 756–62.

187 **buried among all the good news:** Watkins, *The Estrogen Elixir.*

188 **headlines shocked, scared, enraged:** R. D. Langer, "The Evidence Base for HRT: What Can We Believe?" *Climacteric* 20, no. 2 (2017): 91–96.

188 **"The goal of the WHI":** Dr. JoAnn Manson, author interview.

189 **Prescriptions fell by nearly half:** Krieger et al., "Hormone Replacement Therapy, Cancer, Controversies, and Women's Health."

189 **no difference in death rates:** J. E. Manson et al. for the WHI Investigators, "Menopausal Hormone Therapy and Long-Term All-Cause and Cause-Specific Mortality: The Women's Health Initiative Randomized

Trials," *Journal of the American Medical Association* 318, no. 10 (2017): 927–38.

189 **Manson told Reuters:** Lisa Rapaport, "Menopause Hormone Not Linked to Premature Death," Reuters Health, September 12, 2017.

190 **In 2010, a contaminated hormone:** Nanette Santoro et al., "Compounded Bioidentical Hormones in Endocrinology Practice: An Endocrine Society Scientific Statement," *Journal of Clinical Endocrinology and Metabolism* 101, no. 4 (2016): 1318–43.

190 **a journalist on assignment for *More*:** Cathryn Jakobson Ramin, "The Hormone Hoax Thousands Fall For," *More*, October 2013, 134–44, 156.

190 **North American Menopause Society:** North American Menopause Society, "The 2017 Hormone Therapy Position Statement of the North American Menopause Society," *Menopause* 24, no. 7 (2017): 728–53.

12. TESTOSTERONE ENDOPRENEURS

John Hoberman, professor at the University of Texas and author of *Testosterone Dreams: Rejuvenation, Aphrodisia, Doping* (California: University of California Press, 2005), was more than helpful as I researched this chapter. I also interviewed several experts in the field who do basic research as well as working with patients: Dr. Alexander Pastuszak; Dr. Shalender Bhasin, director of the research program in men's health, Brigham and Women's Hospital; Dr. Joel Finkelstein, professor of medicine, Massachusetts General Hospital and Harvard Medical School; Dr. Mark Schoenberg, professor and university chair of urology, Montefiore Medical Center and Albert Einstein College of Medicine; Dr. Elizabeth Barrett-Connor, professor of Family Medicine and Public Health, University of California, San Diego; Dr. Frank Lowe, professor of urology, Albert Einstein College of Medicine; Dr. Martin Miner, co-director of the Men's Health Center, Miriam Hospital, Providence, RI, and associate professor of family medicine, Brown University; Dr. Michael Werner, medical director of the Maze Health Clinic; Dr. Thomas Perls, director of the New England Centenarian Study and professor of medicine, Boston University; Dr. Paul Turek, urologist and founder of the Turek Clinics; Hershel Raff, PhD, professor of medicine, surgery, and physiology and director of endocrine research, Medical College of Wisconsin; Dr. Elizabeth Wilson, professor of pediatrics, biochemistry, and biophysics, University of North Carolina; and Dr. James Dupree, assistant professor of urology, University of Michigan. Historical background comes from Arlene Weintraub, *Selling the Fountain of Youth: How the Anti-Aging Industry Made a Disease Out of Getting Old—And Made Billions* (New York: Basic Books: 2010).

194 **The dogs had sex:** Frank A. Beach, "Locks and Beagles," *American Psychologist* 24, no. 11 (1969): 971–89; Benjamin D. Sachs, "In Memoriam: Frank Ambrose Beach," *Psychobiology* 16, no. 4 (1988): 312–14.

195 **"Yes, manhood is chemical":** Paul de Kruif, *The Male Hormone* (New York: Harcourt, Brace, 1945), 107.

195 **others excoriated testosterone therapy:** W. O. Thompson, "Uses and Abuses of the Male Sex Hormone," *Journal of the American Medical Association* 132, no. 4 (1946): 185–88; Blakeslee, "Stimulant Found in Pituitary Powder."

196 **"If the hypothesis were confirmed":** Beach, "Locks and Beagles."

196 **the debate continues:** Andrea Busnelli et al., "'Forever Young'—Testosterone Replacement Therapy: A Blockbuster Drug Despite Flabby Evidence and Broken Promises," *Human Reproduction* 32, no. 4 (2017): 719–24.

197 **Dr. Fred Koch:** Alvaro Morales, "The Long and Tortuous History of the Discovery of Testosterone and Its Clinical Application," *Journal of Sexual Medicine* 10, no. 4 (2013): 1178–83.

197 **"It is our feeling that until more is known":** T. F. Gallagher and Fred C. Koch, "The Testicular Hormone," *Journal of Biological Chemistry* 84, no. 2 (1929): 495–500.

198 **"So to think of them as growth hormones":** Claudia Dreifus, "A Conversation with—Anne Fausto-Sterling; Exploring What Makes Us Male or Female," *New York Times*, January 2, 2001; Anne Fausto-Sterling, *Sexing the Body* (New York: Basic Books, 2000).

200 **The work was so groundbreaking:** "Science Finds Way to Produce Male Hormone Synthetically," *New York Herald Tribune*, September 16, 1935; "Chemist Produces Potent Hormone," *New York Times*, September 16, 1935; "Testosterone," *Time*, September 23, 1935.

200 **"all the testosterone the world needs":** "Testosterone," *Time*.

201 **direct-to-consumer drug advertisements:** Sarita Metzger and Arthur L. Burnett, "Impact of Recent FDA Ruling on Testosterone Replacement Therapy (TRT)," *Translational Andrology and Urology* 5, no. 6 (2016): 921–26. For an example of the news reports, see Julie Revelant, "10 Warning Signs of Low Testosterone Men Should Never Ignore," *Fox News Health*, July 18, 2016, http://www.foxnews.com/health/2016/07/18/10-warning-signs-low-testosterone-men-should-never-ignore.html.

202 **rebranded the syndrome "Low T":** August Werner, "The Male Climacteric," *Journal of the American Medical Association* 112, no. 15 (1939): 1441–43.

203 **"it's a crappy questionnaire":** Dr. John Morley, author interview.

203 **In a tell-all essay:** Stephen R. Braun, "Promoting 'Low T': A Medical Writer's Perspective," *JAMA Internal Medicine* 173, no. 15 (2013): 1458–60.

203 **ethics started to rattle him:** Stephen Braun, author interview.

203 **All of these tactics:** C. Lee Ventola, "Direct-to-Consumer Pharmaceutical Advertising: Therapeutic or Toxic?" *Pharmacy and Therapeutics* 36, no. 10 (2011): 669–84; Samantha Huo et al., "Treatment of Men for 'Low Testosterone': A Systematic Review," *PLOS ONE* 11, no. 9 (2016): e0162480.

203 **"suddenly it seemed as though the law:** Hoberman, *Testosterone Dreams,* 120.

204 **package inserts:** Metzger and Burnett, "Impact of Recent FDA Ruling."

204 **issued similar guidelines:** Shalender Bhasin et al., "Testosterone Therapy in Men with Androgen Deficiency Syndromes: An Endocrine Society Clinical Practice Guideline," *Journal of Clinical Endocrinology and Metabolism* 95, no. 6 (2010): 2536–59; Frederick Wu et al., "Identification of Late-Onset Hypogonadism in Middle-Aged and Elderly Men," *New England Journal of Medicine* 363, no. 2 (2010): 123–35; G. R. Dohle et al., "Guidelines on Male Hypogonadism," European Association of Urology, 2014, http://uroweb.org/wp-content/uploads/18-Male-Hypogonadism_ LR1.pdf.

204 **Despite the FDA guidelines:** Joseph Scott Gabrielsen et al., "Trends in Testosterone Prescription and Public Health Concerns," *Urologic Clinics of North America* 43, no. 2 (2016): 261–71; Katherine Margo and Robert Winn, "Testosterone Treatments: Why, When, and How?" *American Family Physician* 73, no. 9 (2006): 1591–98.

204 **"a mass uncontrolled experiment":** L. M. Schwartz and S. Woloshin, "Low 'T' as in 'Template': How to Sell Disease," *JAMA Internal Medicine* 173, no. 15 (2013): 1460–62.

204 **Testosterone levels fluctuate:** W. J. Bremner et al., "Loss of Circadian Rhythmicity in Blood Testosterone Levels with Aging in Normal Men," *Journal of Clinical Endocrinology and Metabolism* 56, no. 6 (1983): 1278–81.

205 **Testosterone increases muscle mass:** Fred Sattler et al., "Testosterone and Growth Hormone Improve Body Composition and Muscle Performance in Older Men," *Journal of Clinical Endocrinology and Metabolism* 94, no. 6 (2009): 1991–2001.

205 **"any exogenous testosterone ":** Dr. Alexander Pastuszak, author interview.

205 **not a reliable contraceptive:** A. M. Matsumoto, "Effects of Chronic Testosterone Administration in Normal Men: Safety and Efficacy of High Dosage Testosterone and Parallel Dose-Dependent Suppression of Luteinizing Hormone, Follicle-Stimulating Hormone, and Sperm Production," *Journal of Clinical Endocrinology and Metabolism* 70, no. 1 (1990): 282–87.

205 **more likely to lose belly fat:** L. Frederiksen et al., "Testosterone Therapy Decreases Subcutaneous Fat and Adiponectin in Aging Men," *European Journal of Endocrinology* 166, no. 3 (2012): 469–76.

205 **cardiovascular problems:** Shehzad Basaria et al., "Adverse Events Associated with Testosterone Administration," *New England Journal of Medicine* 363, no. 2 (2010): 109–22; Shehzad Basaria et al., "Effects of Testosterone Administration for 3 Years on Subclinical Atherosclerosis Progression in Older Men with Low or Low-Normal Testosterone Levels: A Randomized Clinical Trial," *Journal of the American Medical Association* 314, no. 6 (2015): 570–81.

205 **Most of the studies:** P. J. Snyder et al., "Effects of Testosterone Treatment in Older Men," *New England Journal of Medicine* 374, no. 7 (2016): 611–24.

205 **giving testosterone to men with normal levels:** Felicitas Buena et al., "Sexual Function Does Not Change when Serum Testosterone Levels Are Pharmacologically Varied within the Normal Male Range," *Fertility and Sterility* 59, no. 5 (1993): 1118–23; Christina Wang et al., "Transdermal Testosterone Gel Improves Sexual Function, Mood, Muscle Strength, and Body Composition Parameters in Hypogonadal Men," *Journal of Clinical Endocrinology and Metabolism* 85, no. 8 (2000): 2839–53.

206 **"You don't see":** Dr. Shalender Bhasin, author interview.

206 **didn't enhance their female-seeking tendencies:** Darius Paduch et al., "Testosterone Replacement in Androgen-Deficient Men With Ejaculatory Dysfunction: A Randomized Controlled Trial," *Journal of Clinical Endocrinology and Metabolism* 100, no. 8 (2015): 2956–62; Snyder et al., "Effects of Testosterone Treatment in Older Men."

206 **no better than placebo:** S. M. Resnick et al., "Testosterone Treatment and Cognitive Function in Older Men With Low Testosterone and Age-Associated Memory Impairment," *Journal of the American Medical Association* 317, no. 7 (2017): 717–27.

206 **Dr. Joel Finkelstein:** Joel S. Finkelstein et al., "Gonadal Steroids and Body Composition, Strength, and Sexual Function in Men," *New England Journal of Medicine* 369, no. 11 (2013): 1011–22.

207 **PATH:** Partnership for the Accurate Testing of Hormones, "PATH Fact Sheet: The Importance of Accurate Hormone Tests," Endocrine Society, Washington DC, 2017.

207 **This harks back to the notion:** Eder, "The Birth of Gender," 83.

207 **"For some reason":** Dr. Mohit Khera, author interview. Mohit Khera et al., "Adult-Onset Hypogonadism," *Mayo Clinic Proceedings* 91, no. 7 (2016): 908–26. Mohit Khera, "Male Hormones and Men's Quality of Life," *Current Opinion in Urology* 26, no. 2 (2016): 152–57.

208 **articles slamming the Low-T industry:** Natasha Singer, "Selling That New-Man Feeling," *New York Times*, November 23, 2013; Sky Chadde, "How the Low T Industry Is Cashing in on Dubious, and Perhaps Dangerous, Science," *Dallas Observer*, November 12, 2014; Sarah Varney, "Testosterone, The Biggest Men's Health Craze Since Viagra, May Be Risky," *Shots: Health News from NPR*, April 28, 2014, http://www.npr.org/sections/health-shots/2014/04/28/305658501/prescription-testosterone-the-biggest-men-s-health-craze-since-viagra-may-be-ris.

210 **To become board-certified:** Rona Schwarzberg, educational advisor at the American Academy of Anti-Aging Medicine, author interview. https://www.a4m.com/certification-in-metabolic-and-nutritional-medicine.html.

210 **"these capitalists have constructed":** Weintraub, *Selling the Fountain of Youth*.

211 **two opposing opinion pieces:** Adriane Fugh-Berman, "Should Family Physicians Screen for Testosterone Deficiency in Men? No: Screening May Be
Harmful, and Benefits Are Unproven" *American Family Physician* 91, no.
4 (2015): 227–28; J. J. Heidelbaugh, "Should Family Physicians Screen for
Testosterone Deficiency in Men? Yes: Screening for Testosterone Deficiency Is Worthwhile for Most Older Men," *American Family Physician* 91,
no. 4 (2015): 220–21.

211 **more than 5,000 men who claim:** Arlene Weintraub, "What's Next for the
Thousands of Angry Men Suing Over Testosterone?," *Forbes* online, April 6,
2015, http://www.forbes.com/sites/arleneweintraub/2015/04/06/whats-next
-for-the-thousands-of-angry-men-suing-over-testosterone/#7cd2401f4833;
Arlene Weintraub, "AbbVie Challenges Fairness of Upcoming Testosterone Trials," *Forbes* online, August 17, 2015, https://www.forbes.com/sites/
arleneweintraub/2015/08/17/abbvie-challenges-fairness-of-upcoming
-testosterone-trials/2b39e0113901; Arlene Weintraub, "Testosterone Suits
Soar Past 2,500 as Legal Milestone Looms for AbbVie," *Forbes* online, October 30,
2015, http://www.forbes.com/sites/arleneweintraub/2015/10/30/testosterone
-suits-soar-past-2500-as-legal-milestone-looms-for-abbvie/57c9501b1199;
Arlene Weintraub, "Why All Those Testosterone Ads Constitute Disease
Mongering," *Forbes* online, March 24, 2015, http://www.forbes.com/sites/
arleneweintraub/2015/03/24/why-all-those-testosterone-ads-constitute
-disease-mongering/#629d9d585853.

211 **On July 24, a federal jury:** Lisa Schencker, "AbbVie Must Pay $150 Mil-
lion over Testosterone Drug, Jury Decides," *Chicago Tribune*, July 24, 2017,
http://www.chicagotribune.com/business/ct-abbvie-androgel-decision
-0725-biz-20170724-story.html.

212 **"I was just shocked":** Dr. Peter Klopfer, author interview.

13. OXYTOCIN: THAT LOVIN' FEELING

This chapter is based on interviews with Dr. Peter Klopfer, professor emeritus
of biology, Duke University; Dr. Cort Pedersen, professor of psychiatry and
neurobiology, University of North Carolina; and Dr. Robert Froemke, associate professor of neuroscience, New York University, whose laboratory I visited.
Gideon Nave, PhD, assistant professor of marketing, Wharton School, University of Pennsylvania, helped me sort through the statistics. Dr. Steve Chang,
assistant professor of psychology and neurobiology, Yale University, talked to
me about his work with monkeys and oxytocin; Dr. Jennifer Bartz, associate professor of psychology, McGill University, spoke with me about oxytocin
and autism. I also interviewed Dr. Michael Platt, professor of anthropology,
University of Pennsylvania, and Dr. James Higham, principal investigator in
primate reproductive ecology and evolution, New York University.

216 **"I was, frankly, not at all attracted":** John G. Simmons, "Henry Dale: Discovering the First Neurotransmitter," chapter in *Doctors and Discoveries: Lives that Created Today's Medicine* (Boston: Houghton Mifflin Harcourt, 2002), 238–427.

217 **and earned Dale a Nobel Prize:** H. O. Schild, "Dale and the Development of Pharmacology: Lecture given at Sir Henry Dale Centennial Symposium, Cambridge, 17–19 September 1975," *British Journal of Pharmacology* 120, Suppl. 1 (1997): 504–8; www.nobelprize.org/nobel_prizes/medicine/laureates/1936/dale-bio.html.

217 **"The pressor principle":** Sir Henry Dale, "On Some Physiological Aspects of Ergot," *Journal of Physiology* 34, no. 3 (1906):163–206.

218 **link to breast milk:** Mavis Gunther, "The Posterior Pituitary and Labour," letter to the editor, *British Medical Journal* 1948, no. 1: 567.

219 **a goat study by Peter Klopfer:** Peter H. Klopfer, "Mother Love: What Turns It On? Studies of Maternal Arousal and Attachment in Ungulates May Have Implications for Man," *American Scientist* 59, no. 4 (1971): 404–7.

220 **expand his mother-newborn studies:** David Gubernick and Peter H. Klopfer, eds., *Parental Care in Mammals* (New York: Plenum Press, 1981).

220 **a balloon-like contraption:** Klopfer, "Mother Love."

220 **University of Cambridge team:** E. B. Keverne et al., "Vaginal Stimulation: An Important Determinant of Maternal Bonding in Sheep," *Science* 219, no. 4580 (1983): 81–83.

221 **an oxytocin expert:** M. L. Boccia et al., "Immunohistochemical Localization of Oxytocin Receptors in Human Brain," *Neuroscience* 253 (2013): 155–64; Cort Pedersen et al., "Intranasal Oxytocin Blocks Alcohol Withdrawal in Human Subjects," *Alcoholism: Clinical and Experimental Research* 37, no. 3 (2013): 484–89; Cort A. Pedersen, *Oxytocin in Maternal, Sexual and Social Behaviors* (New York: New York Academy of Sciences, 1992).

221 **as if they were trying to breastfeed:** Dr. Cort Pedersen, author interview.

222 **Further experiments:** C. A. Pedersen et al., "Oxytocin Antiserum Delays Onset of Ovarian Steroid-Induced Maternal Behavior," *Neuropeptides* 6, no. 2 (1985): 175–82; E. van Leengoed, E. Kerker, and H. H. Swanson, "Inhibition of Postpartum Maternal Behavior in the Rat by Injecting an Oxytocin Antagonist into the Cerebral Ventricles," *Journal of Endocrinology* 112, no. 2 (1987): 275–82.

222 **remained aloof:** Pedersen, *Oxytocin in Maternal, Sexual and Social Behaviors.*

222 **but not actual sexual performance:** D. M. Witt et al., "Enhanced Social Interactions in Rats Following Chronic, Centrally Infused Oxytocin," *Pharmacology Biochemistry and Behavior* 43, no. 3 (1992): 855–61.

223 **Sue Carter:** C. S. Carter and L. L. Getz, "Monogamy and the Prairie Vole," *Scientific American* 268, no. 6 (1993): 100–6.

223 **Stanford University scientists:** M. S. Carmichael et al., "Plasma Oxytocin

Increases in the Human Sexual Response," *Journal of Clinical Endocrinology and Metabolism* 64, no. 1 (1987): 27–31.

223 **oxytocin prompted feelings of peacefulness:** C. S. Carter, *Hormones and Sexual Behavior* (Stroudsburg, PA: Dowden, Hutchinson & Ross, 1974).

223 **Other studies suggest:** A. S. McNeilly et al., "Release of Oxytocin and Prolactin in Response to Suckling," *British Medical Journal (Clinical Research Edition)* 286, no. 6361 (1983): 257–59.

223 **The really dramatic experiment:** M. M. Kosfeld et al., "Oxytocin Increases Trust in Humans," *Nature* 435, no. 7042 (2005): 673–76.

224 **"the moral molecule":** P. J. Zak, *The Moral Molecule: How Trust Works* (New York: Plume, 2013); V. Noot, *35 Tips for a Happy Brain: How to Boost Your Oxytocin, Dopamine, Endorphins, and Serotonin* (CreateSpace, 2015).

225 **Zak once blogged:** Paul. J. Zak, "Why Love Sometimes Sucks," *Huffington Post*, December 5, 2012, http://www.huffingtonpost.com/paul-j-zak/why-love-sometimes-sucks_b_1504253.html.

225 **"What we're left with":** Ed Yong, "The Weak Science Behind the Wrongly Named Moral Molecule," *Atlantic*, November 13, 2015.

225 **"It's a great story":** Gideon Nave, author interview.

226 **"It is pathetic":** Hans Lisser to Dr. Cushing, July 19, 1921.

227 **careful studies of oxytocin:** B. J. Marlin et al., "Oxytocin Enables Maternal Behaviour by Balancing Cortical Inhibition," *Nature* 520, no. 7548 (2015): 499–504; Helen Shen, "Neuroscience: The Hard Science of Oxytocin," *Nature* 522, no. 7557 (2015): 410–12; Marina Eliava et al., "A New Population of Parvocellular Oxytocin Neurons Controlling Magnocellular Neuron Activity and Inflammatory Pain Processing," *Neuron* 89, no. 6 (2016): 1291–1304.

227 **zeroing in on the precise location:** Michael Numan and Larry J. Young, "Neural Mechanisms of Mother–Infant Bonding and Pair Bonding: Similarities, Differences, and Broader Implications," *Hormones and Behavior* 77 (2016): 98–112; Shen, "Neuroscience: The Hard Science of Oxytocin."

227 **Froemke's work built on:** McNeilly et al., "Release of Oxytocin and Prolactin in Response to Suckling."

228 **tap its potential:** Robert C. Liu, "Sensory Systems: The Yin and Yang of Cortical Oxytocin," *Nature* 520, no. 7548 (2015): 444–45.

14. TRANSITIONING

This chapter is based on interviews with Mel Wymore and shaped by discussions with others in the transgender community. I also interviewed clinicians including Dr. Joshua Safer, Dr. Anisha Patel, Dr. Susan Boulware, Leslie Henderson, PhD, and Dr. Jack Turban. Dr. Howard W. Jones, Jr., and Claude Migeon provided details on the early days of transgender therapy. For background I consulted Joanne Meyerowitz, *How Sex Changed: A History of Transsexuality in the United States* (Cambridge, MA: Harvard University Press, 2004) and several memoirs: Jenny Boylan, *She's Not There: A Life in Two Genders* (New York: Broadway Books, 2013); Amy Ellis Nutt, *Becoming Nicole: The Transformation of an American Family* (New York: Random House, 2015); Julia Serrano, *Whipping Girl: A Transsexual Woman on Sexism and the Scapegoating of Femininity* (Berkeley, CA: Seal Press: 2007); Pagan Kennedy, *The First Man-Made Man* (New York: Bloomsbury, 2007); Christine Jorgensen, *Christine Jorgensen: A Personal Autobiography* (New York: Bantam, 1968), and Andrew Solomon, "Transgender," chapter 11 in *Far From the Tree* (New York: Scribner, 2012), 599–676.

231 **between 0.3 and 0.6 percent of people:** Sari L. Reisner et al., "Global Health Burden and Needs of Transgender Populations: A Review," *Lancet* 388, no. 10042 (2016): 412–36.

231 **1.4 million transgender Americans:** https://williamsinstitute.law.ucla.edu/ wp-content/uploads/How-Many-Adults-Identify-as-Transgender-in-the -United-States.pdf.

231 **spate of media:** As well as the books mentioned above: Deirdre W. McCloskey, *Crossing: A Memoir* (Chicago: University of Chicago Press, 1999); Max Wolf Valerio, *The Testosterone Files* (Berkeley, CA: Seal Press: 2006); Jamison Green, *Becoming a Visible Man* (Nashville: Vanderbilt University Press, 2004). Documentaries include *Gender Revolution: A Journey with Katie Couric*, National Geographic, 2017. Articles include Rachel Rabkin Peachman, "Raising a Transgender Child," *New York Times Magazine*, January 31, 2017, and Hannah Rosin, "A Boy's Life," *Atlantic*, November 2008. See also Jill Soloway's television series *Transparent*.

231 **The rise of plastic surgery in the early twentieth century:** Felix Abraham, "Genitalumwandlungen an zwei männlichen Transvestiten," *Zeitschrift für Sexualwissenschaft und Sexualpolitik* 18 (1931): 223–26, describes operations at the Institute for Sexual Science, founded by Magnus Hirchfield, and described in Meyerowitz, *How Sex Changed*. The story of Danish painter Lili Elbe was told in the 2015 film *The Danish Girl*.

231 **The big difference between then and now:** Wylie C. Hembree et al., "Endocrine Treatment of Gender-Dysphoric/Gender-Incongruent Persons: An Endocrine Society Clinical Practice Guideline," *Journal of Clinical Endocrinology and Metabolism* 102, no. 11 (2017): 3869–903.

232 **"Ex-GI Becomes Blonde Beauty":** *New York Daily News*, December 1, 1952.

233 **"Could the transition":** Jorgensen, *Christine Jorgensen*, 72.

234 **front-page story:** "Surgery Makes Him a Woman," *Chicago Daily Tribune*, December 1, 1952.

234 **"I mean are you interested":** United Press, "My Dear, Did You Hear About My Operation?" *Austin Statesman*, December 2, 1952.

234 **"I somehow skyrocketed":** Jorgensen, *Christine Jorgensen*, 218.

234 ***The Transsexual Phenomenon:*** Dr. Harry Benjamin, *The Transsexual Phenomenon* (New York: Julian Press, 1966).

235 **"it is a possibility":** Harry Benjamin, introduction to Jorgensen, *Christine Jorgensen*, x.

236 **Money's central tenets:** see chapter 7, p. 114.

236 **Nowadays, scientists:** Leslie Henderson, PhD, and Dr. Joshua Safer, author interviews. See also Margaret M. McCarthy and A. P. Arnold, "Reframing Sexual Differentiation of the Brain," *Nature Neuroscience* 14, no. 6 (2011): 677–83; S. A. Berenbaum and A. M. Beltz, "Sexual Differentiation of Human Behavior: Effects of Prenatal and Pubertal Organizational Hormones," *Frontiers in Neuroendocrinology* 32, no. 2 (2011): 183–200; I. Savic, A. Garcia-Falgueras, and D. F. Swaab, "Sexual Differentiation of the Human Brain in Relation to Gender Identity and Sexual Orientation," *Progress in Brain Research* 186 (2010): 41-62; and Elke Stefanie Smith et al., "The Transsexual Brain—A Review of Findings on the Neural Basis of Transsexualism," *Neuroscience and Biobehavioral Reviews* 59 (2015): 251–66.

236 **classic 1959 study:** Charles Phoenix et al., "Organizing Action of Prenatally Administered Testosterone Propionate on the Tissues Mediating Mating Behavior in the Female Guinea Pig," *Endocrinology* 65 (1959): 369–82, reprinted in *Hormonal Behavior* 55, no. 5 (2009): 566.

237 **"People, even some scientists":** Leslie Henderson, PhD, author interview.

237 **Further rodent studies:** For a thorough recent review see Margaret M. McCarthy, "Multifaceted Origins of Sex Differences in the Brain," *Philosophical Transactions of the Royal Society B* 371, no. 1688 (2016).

238 **"It seems pretty clear":** Dr. Joshua Safer, author interview.

242 **Doctors keep an eye on other hormone side effects:** Ibid. On the influence of hormone treatment on serotonin receptors, which may influence depression, see G. S. Kranz et al., "High-Dose Testosterone Treatment Increases Serotonin Transporter Binding in Transgender People," *Biological Psychiatry* 78, no. 8 (2015): 525–33. On the impact of hormone therapy on transgender patients, see Cécile A. Unger, "Hormone Therapy for Transgender Patients," *Translational Andrology and Urology* 5, no. 6 (2016): 877–84.

242 **The most recent Endocrine Society guidelines:** Hembree et al., "Endocrine Treatment of Gender-Dysphoric/Gender-Incongruent Persons."

245 **More than 40 percent:** Ibid.; Ann P. Haas, PhD, et al., "Suicide Attempts Among Transgender and Gender Non-Conforming Adults," Williams Institute, https://williamsinstitute.law.ucla.edu/wp-content/uploads/AFSP-Williams-Suicide-Report-Final.pdf.

15. INSATIABLE: THE HYPOTHALAMUS AND OBESITY

This chapter is based on extensive interviews with Karen Snizek and interviews with Dr. Rudolph L. Leibel, professor of pediatrics and medicine at the Institute of Human Nutrition, Columbia University College of Physicians and Surgeons;

Dr. Jeffrey M. Friedman, director of the Starr Center for Human Genetics, Rockefeller University; and Sir Stephen O'Rahilly, Professor of Clinical Biochemistry and Medicine, University of Cambridge, and his colleague I. Sadaf Farooqi, a specialist in metablism and medicine, who are at the forefront of drug research. I also interviewd Dr. Gerald Schulman, professor of cellular and molecular physiology, Yale University; Dr. Frank Greenway, medical director of the outpatient clinic, Pennington Biomedical Research, Baton Rouge, LA; and Dr. Jennifer Miller of the University of Florida.

247   **Rats don't vomit:** Ruth B. S. Harris, "Is Leptin the Parabiotic 'Satiety' Factor? Past and Present Interpretations," *Appetite* 61, no. 1 (2013): 111–18. For further information on rats and vomiting see Charles C. Horn et al., "Why Can't Rodents Vomit? A Comparative Behavioral, Anatomical, and Physiological Study," *PLOS One*, April 10, 2013.

248   **a stunningly simple yet quirky experiment:** G. R. Hervey, "The Effects of Lesions in the Hypothalamus in Parabiotic Rats," *Journal of Physiology* 145, no. 2 (1959): 336–52; G. R. Hervey, "Control of Appetite: Personal and Departmental Recollections," *Appetite* 61, no. 1 (2013): 100–10.

249   **hunt for the elusive substance:** Ellen Rupel Shell, *The Hungry Gene: The Inside Story of the Obesity Epidemic* (New York: Grove Press, 2003); "Douglas Coleman: Obituary," *Daily Telegraph*, April 17, 2014.

249   **cholecystokinin:** E. Straus and R. S. Yalow, "Cholecystokinin in the Brains of Obese and Nonobese Mice," *Science* 203, no. 4375 (1979): 68–69.

249   **proved her wrong:** B. S. Schneider et al., "Brain Cholecystokinin and Nutritional Status in Rats and Mice," *Journal of Clinical Investigation* 64, no. 5 (1979): 1348–56.

250   **In 1994, inspired by Coleman's work:** Y. Zhang et al., "Positional Cloning of the Mouse Obese Gene and Its Human Homologue," *Nature* 372, no. 6505 (1994): 425–32.

251   **"We didn't appreciate":** Dr. Rudy Leibel, author interview.

251   **sparked a sensation:** Tom Wilkie, "Genes, Not Greed, Make You Fat," *Independent*, December 1, 1994; Natalie Angier, "Researchers Link Obesity in Humans to Flaw in a Gene," *New York Times*, December 1, 1994.

251   **"so when leptin is low":** Dr. Jeffrey Friedman, author interview.

253   **Thanks to research:** L. G. Hersoug et al., "A Proposed Potential Role for Increasing Atmospheric CO_2 as a Promoter of Weight Gain and Obesity," *Nutrition and Diabetes* 2, no. 3 (2012): e31.

253   **Endocrinologists are joining forces:** Anthony P. Coll et al., "The Hormonal Control of Food Intake," *Cell* 129, no. 2 (2007): 251–62.

253   **germs may give us a tendency to gain weight:** Dorien Reijnders et al., "Effects of Gut Microbiota Manipulation by Antibiotics on Host Metabolism in Obese Humans: A Randomized Double-Blind Placebo-Controlled Trial," *Cell Metabolism* 24, no. 1 (2016): 63–74.

253 **thorough understanding of hunger:** Ilseung Cho and Martin J. Blaser, "The Human Microbiome: At the Interface of Health and Disease," *Nature Reviews Genetics* 13, no. 4 (2012): 260–70; Torsten P. M. Scheithauer et al., "Causality of Small and Large Intestinal Microbiota in Weight Regulation and Insulin Resistance," *Molecular Metabolism* 5, no. 9 (2016): 759–70.

253 **air pollution:** Y. Wei et al., "Chronic Exposure to Air Pollution Particles Increases the Risk of Obesity and Metabolic Syndrome: Findings from a Natural Experiment in Beijing," *FASEB Journal* 30, no. 6 (2016): 2115–22.

253 **industrial chemicals:** G. Muscogiuri et al., "Obesogenic Endocrine Disruptors and Obesity: Myths and Truths," *Archives of Toxicology*, October 3, 2017, https://doi.org/10.1007/s00204-017-2071-1; K. A. Thayer, J. J. Heindel, J. R. Bucher, and M. A. Gallo, "Role of Environmental Chemicals in Diabetes and Obesity: A National Toxicology Program Workshop Review," *Environmental Health Perspectives* 120 (2012): 779–89.

254 **weight loss surgery:** Valentina Tremaroli et al., "Roux-en-Y Gastric Bypass and Vertical Banded Gastroplasty Induce Long-Term Changes on the Human Gut Microbiome Contributing to Fat Mass Regulation," *Cell Metabolism* 22, no. 2 (2015): 228–38.

254 **less efficient calorie burners:** Wendee Holtcamp, "Obesogens: An Environmental Link to Obesity," *Environmental Health Perspectives* 120, no. 2 (2012): a62–a68; David Epstein, "Do These Chemicals Make Me Look Fat?" *ProPublica*, October 11, 2013; Jerrold Heindel, "Endocrine Disruptors and the Obesity Epidemic," *Toxicological Sciences* 76, no. 2 (2003): 247–49.

254 **feeding the obesity epidemic:** Yann C. Klimentidis et al., "Canaries in the Coal Mine: A Cross-Species Analysis of the Plurality of Obesity Epidemics," *Proceedings of the Royal Society B: Biological Sciences*, 2010, doi: 10.1098/rspb.2010.1890.

Index

Note: Page numbers in italics indicate figures; 'n' indicates note at bottom of page.

AbbVie, 207, 211
achondroplasia, 127
acromegaly, 44–45, 53, 136
Act to Amend the Law Relating to
    Cruelty to Animals (England),
    23–24, 25, 27–28
ACTH (adrenocorticotropic hormone),
    47
A.D.A.M. questionnaire (androgen defi-
    ciency in the aging male), 202–3
Addison, Thomas, 12
Addison's disease, xii–xiii, xv–xvi, 12,
    258
"adrenal centered," 60
adrenal glands, 8, 10, 12, 18, 30, 43–44,
    60, 111–12, 253, 258
adrenal hormones, 108. *See also*
    adrenaline
adrenaline, 12, 58
    in ergot experiments, 216–17
    hot flashes and, 180
    organotherapy and, 62
"adrenal inferior," 60

advice books, xiv
age-related low testosterone, 202–4
aging, testosterone and, 196–97
Albert Einstein Hospital, 134
Allen, Edgar, 92
Alzheimer's disease, 177, 183, 187, 209
American Academy of Anti-Aging
    Medicine, 208–10
American Academy of Family Physi-
    cians, 209–10
American Association for the Advance-
    ment of Science, 57
American College of Obstetricians and
    Gynecologists, 190
American College of Physicians, 187
American Society for Reproductive
    Medicine, 190
amino acids, 7
Aminoff, Michael, 162
amphetamines, 200
amygdala, 237
AndroGel, 201–2
androgen insensitivity, 123

androgens, 111–12, 200
animal glands, 200, 208
animal research, 29, 111, 153, 197, 198
    Cushing and, 40–41
    with ergot, 216–17
    gender identity and, 236–37
    growth hormones and, 128–29, 136
    maternal bonding and, 219–20
    obesity and, 247–49
    oxytocin and, 217–23
    restrictions on, 23–24, 27–28
    sex and, 237
    Steinach and, 77–84
    testosterone and, 194–98, 212–13
    vivisection, 14–19, 23–28, 29,
        31–33, 40–41
animal rights activists, 14–19, 23,
    25–28, 31–33
animals, as source of hormones,
    136–37, 200, 208
animal testing, pregnancy tests and,
    93–95, 94n
animal vivisection, 14–19, 23–28, 29,
    31–33, 40–41. See also animal
    research
anterior pituitary, 39, 41, 45, 47
anti-aging doctors, 208
anti-androgens, 242, 243
antibiotics, 108, 163, 253
antibodies, hormones and, 154–57
anti-Mullerian hormone, 111
Anti-Pituitary Tumor Club, 47
antivivisectionists, 14–19, 23, 25–28,
    31–33
Anton W., 84
anxiety, 180, 228
appetite, 248–56
Aschheim, Selmar, 94–96, 99
Assocation for the Study of Internal
    Secretions, 57, 257
Astor, William Vincent, 38
athletes, testosterone and, 200
autism, 227–28
"autocoid," 29–30
Ayerst, 185
A–Z test (Aschheim–Zondek test),
    93–96

bacteria, 253
Balaban, Al, 126–27, 130–35, 138–42,
    146, 169
Balaban, Barbara, 126–27, 130–35,
    137–42, 146, 164, 169
Balaban, Jeffrey, 126–27, 129,
    130–35, 137–42, 146, 156, 169,
    172–73
Banting, Frederick, 57
Baratz, Arlene, 122–23
Baratz, Katie, 122–23
basophil adenoma, 46
Battersea Park, 32
Bayliss, William, 15, 18–25, 27–28, 33,
    218
Bazedoxifene, 192
Beach, Frank, 194–96, 195n, 211–13
behavior, 8
    behaviorism, 64–65
    pseudo-homosexual behavior, 83
    sexual behavior, 237
behavioral endocrinology, 212
behaviorism, 64–65
Benjamin, Harry, 234–35
Berman, Louis, 59–65, 72, 79
    reception of, 62–64
    Sing-Sing research, 65–66
Bernard, Claude, xii–xiii
Berson, Solomon, 152–58
Berthold, Arnold, 10–12, 29, 80, 197,
    218
Best, Charles, 57
Bevan, Rosie (Wilmot), 44–45, 45n
Bhasin, Shalender, 206
binary gender system, 125
biochemistry, advances in, xiv–xv
biotech companies, 145–46
birth control pills, xv, 182, 185–86
Blakeslee, Alton, 193
Blavatsky, Helena, 69
Bliss, Michael, 38
Blizzard, Robert, 143–44, 166–67,
    169–70, 172–73
blood clots
    hormone replacement therapy and,
        188–89
    testosterone and, 211

blood sugar, hormone therapy and, 160–61
Boston University, 238
Bowman, Karl, 66, 69
brain, impact of hormones on, 66
brain cells
  brain cell messengers, 9
  hot flashes and, 181–83
  hunger messages between, 254
brain damage, 69
brain imaging, gender identity and, 237–38
brains, pickled, 34–53
"Brain Society," 49
brain stimulation, hormones and, 80–81
brain tumors, 34, 37, 38, 46
Braun, Stephen, 203
breast cancer, hormone replacement therapy and, 188–89, 191
breastfeeding, 223
Brinkley, John, 74
Bronx Veterans Administration, 152–53
Brown, Paul, 167–68
Brown Dog Affair, 14–17, 23–25, 26, 217n
Brown Dog Society, 14–17, 32–33
Brown Dog statue, 14–17, *16*, 26, 31–33, *32*
Browning, Robert, 60
Brown-Séquard, Édouard, 75
Bruce, Tara, 50
Bruckner, William, 40
Buck v. Bell, 64
bulbus glandis, 195
bulls' testicles, 197
Butenandt, Adolf, 198–200
Byrne, Charles, 42n

cadavers, study of, 41–42. *See also* Grey, Blanche
calcium levels, 60
cancer, hormone replacement therapy and, 183, 187, 188–89, 191. *See also specific kinds of cancer*
Cannon, Walter, 58
Carney complex, 53
Carter, Sue, 223

Caverly, John R., 70
Centers for Disease Control and Prevention, 171
cervical cancer, 175–76
"chalone," 30
Chase, Cheryl, 122, 123–24. *See also* Laurent, Bo; Sullivan, Bonnie; Sullivan, Brian Arthur
chemical messengers, 9
chemistry, 8–9
childbirth, 47, 218–20
childhood, prolonged human, 127–28
children. *See also* infants
  hormone therapy and, 242–44
  intersex, 102–25, 236
  thyroid hormone deficiency in, 148–49
cholecystokinin (CCK), 249–50
cholesterol, 187, 200–201
chorionic gonadotropin, 99
chromosomes, sexual identity and, 114
Ciba, 200–201
Clerc, Laurent, 123–24
clitoris, 106, 119
  amputated, 116, 120–21
  large, 82, 106, 115, 116, 120–21
cognition, testosterone and, 206
Colapinto, John, 117–18
Coleman, Douglas, 249, 250
Coleridge, Stephen, 26–28
colon cancer, 187
Columbia Presbyterian Hospital, 105, 118–19
Columbia University, 104–5, 152, 254
compounded hormones, 189–91
Compounding Quality Act, 191
conformity, 125
congenital adrenal hyperplasia (CAH), 111–13
contraceptive pill. *See* birth control pills
control, 256
conversion therapy, 113–14
cortisol, 43–44, 111–12
cortisol pathway, blocked, 111
cortisone, 112–13

cortisone pills, 253
crapie, 164
"cretinism," 148–49
Creutzfeldt-Jakob disease (CJD), 162–73, *165*
criminality, 54–71
Cromwell, Kate, 36
Curling, Thomas Blizard, 12
Cushing, Harvey, 34–48, 45*n*, 58, 67, 74, 79, 90–91, 135, 217, 226–27, 257–59
    animal studies and, 40–41
    begins working at Yale, 47
    daughters of, 37–38
    death of, 48
    experiments by, 40–41
    family of, 36–38
    fascination with endocrinology, 39
    as founder of neurosurgery, 38
    glass-plate negatives of his patients, 49–50
    on growth hormones, 128*n*
    hormone studies by, 39–41
    letter to editors of *Time* magazine, 44–45
    operating technique of, 38–39
    pituitary gland and, 39–40
    retirement from Harvard, 47–48
    theory of pituitary gland, 39–40, 45, 69
Cushing Brain Tumor Registry, 34, *37*, 38, 47, 48–53, *52*
Cushing Center, 51–53
Cushing's disease, 43–44
Cushing's syndrome, 43–44

Dagradi, Terry, 50–53
Dale, Henry, 24–25, 216–17, 217*n*
Darrow, Clarence, 55, 66, 68, 70
Darwin, Charles, 9
Davenport, Charles, 63–64
Davis, David, 170
Day, James, 71
"db" mice, 249
defense cells, 9
de Kruif, Paul, 195, 200, 233
dementia praecox, 60

depression, 60
DES, 100–101, 120
Descartes, René, 69
deviance, 56. *See also* criminality
diabetes, xiii, 57, 249
Diamond, Milton, 117–18
Dickinson, Alan, 170–71
difference of sex development (DSD), 109–10
disease, hormone levels and, 204
disorder of sex development (DSD), 109–10
dogs, experiments on, 14–20, 23–28, 31–33, 40–41, 136, 194–96
Doisy, Edward, 92
doping, 200
Dresselhaus, Mildred, 151
drug advertisements, direct-to-consumer, 201–2
Duavee, 192
Duke University, 219–20
du Vigneaud, Vincent, 218
dwarfism, 126–47
dyspituitarism, 43

Edwards, Robert, 89
EEGs, 162
Eisenhardt, Louise, 47
Elbe, Lili, 230–31
electrolytes, 22
embryos, hormones and, 93, 110
Emory University, 142–43, 226–27, 228
emotions, sex and, 80–81
endocrine imbalance, criminality and, 65
endocrine products, 62
Endocrine Society, 57, 190, 242
endocrine status, assessment for, 64
endocrine system, 6–8, 18–20. *See also specific organs*
endocrinology, xii–xiii, xiv, 6–7, 18–20, 22, 47. *See also specific subfields*
    in 1920s, xiv–xv, 56–57, 59, 61, 73–74
    animal experimentation and, 14–27, 33

birth of, 75
Cushing's fascination with, 39
deviance and, 56 (*see also* criminality)
early, 14–32
emergence of, 8–11, 12
enthusiasm for, 57
as religion, 64
reproductive, 89–101
skepticism and, 226–27
traditional certification, 209
endometrial cancer, 175, 190
energy, 206, 248–56
Engle, Earl, 95
Enovid, 185
epinephrine suppositories (adrenaline),
62
ergot, 216–17
estriol, 191
estrogen, xiv–xv, 47, 116, 185–86
androgen insensitivity and, 123
blocking of, 241
fat and, 207
genitalia and, 108
gestation and, 115
heart health and, 187
hormone replacement therapy and,
186–89, 191–92
hot flashes and, 175–77, 180, 182, 184
isolation of, xiv, 57
libido and, 207
maternal bonding and, 220–21
menopause and, 175–76
postmenopausal declines in, 22
pure, 184
purification of, 92
synthetic, 100–101
testosterone and, 207
uterine cancer and, 186
estrogen receptors, 182
estrogen therapy, 197. *See also* hormone replacement therapy
estrogen pills, 185, 233–34
estrogen-progesterone pills, 186–90,
191–92
side effects of, 242
for transgender women, 242–44

eugenics, 64
eunuchs, 58
Evans, Herbert, 135–36, 137

fat. *See also* obesity
estrogen and, 207
fat distribution, 242
testosterone and, 205
Fat Bride, 1–13, 247
Fausto-Sterling, Anne, 121, 198
F-cells, 83
Fehr, Ernst, 225
Feldman, Trudi, 71
femaleness, 92–93, 111
fertility business, 89
fertility treatments, 99, 158
Finkelstein, Joel, 206, 208
Fishbein, Morris, 85
Fisher, Helen E., 176
5-alpha-reductase type 1 and type 2
enzymes, 112
Foley, Thomas, 158–59
folk remedies, 216–17
Foster, James Whitney, 38
Franks, Bobby, 55, 70
Freedman, Robert, 178–79, 181, 192
Freud, Sigmund, 73
Freudian analysis, 68
Friedman, Jeffrey, 250–52, 256
Froemke, Robert, 227
"fuckology," 113
Fugh-Berman, Adriane, 211
fungal meningitis, 190

Galen, 39
Gallagher, T. F., 197
Garcia, Nancy, 89–101
gastric acid, 22
gender
brain imaging and, 237–38
childrearing and, 115–16
formation through three-step process, 115
gender roles, 115
genitalia and, 245
gestation and, 115
making, 102–25

gender (*continued*)
  malleability of, 115–16
  prenatal hormones and, 115–16
  sexual anatomy and, 237–38
gene hunting, 250
Genentech, 167–68
genitalia. *See also specific organs*
  ambiguous, 83, 114–15
  atypical/ambiguous, 102–25
  gender identity and, 245
  genetic signals and, 111
  morphology of, 115
  sex hormones and, 108
germ theory, 9, 30–31
gestation, gender and, 115
Gey, George Otto, 93, 96–97
Gey, Margaret, 96, 97
gland medicine, as cure-all, 61–62
glands, xiii, xiv, 7–8, 18–20, 30. *See also specific glands*
  animal glands, 200, 208
  chain of command among, 108
  glandular dysfunction, 69–70
globulins, 154
gonadotropins, 47
gonads. *See also* ovaries; testes
  sex-specificity of, 81–84, 88
  sexual identity and, 114–15
Grace–New Haven Hospital, Yale University, 194–95
grehlin, 3
Grey, Blanche, 1–6, 12–13, 247
Grimm, David, 16
growth, sex hormones and, 128. *See also* growth hormones
growth hormones, xiii–xiv, 3, 8, 47, 125, 130, 132–33, 198
  in 1960s, 128–29
  banning of human growth hormone therapy, 167
  contaminated, 160–73
  deficiency in, 160
  extraction of, 142, 146–47
  hormone therapy and, 133–47, 160–73
  human growth hormone, 133–47, 160–73

  measurement of, 156
  sex hormones as, 198
  synthetic, 145–49, 167–68
  timing of release of, 128
G. W. Carnrick Company, 62
gynandromorphs, 110
gynecology, 174–93

Haire, Norman, 87
Hall, Prudence, 214–15, 228–29
Halsted, William, 36
Hamburger, Christian, 233–34
Hampson, Joan, 117
Hardy, Thomas, 26
Hardy, William B., 29
Harrow, Benjamin, 63
Harvard University, 47, 97, 104, 188
Haseltine, Florence, 186–87, 192–93
healing gurus, 76
Healy, Bernadine, 177
hearing, 227
heart, 19
heart disease, 177, 183
  estrogen and, 187
  hormone replacement therapy and, 187–88, 191
  low testosterone and, 208–9
  testosterone therapy and, 205, 211
Hedielbaugh, Joel, 211
height, hormone therapy and, 126–47
HeLa cell line, 96
Hellman, Louis, 97
Henderson, Leslie, 237
Henneman, Philip, 137
herbal remedies, 215
Herder, Wouter de, 45*n*
hermaphroditism, 83, 109, 111, 120, 121
Hermaphroditus, 109
Hertig, Arthur, 97
Hervey, George R., 248–49
heterosexuality, 83, 113–14
Higdon, Hal, 70
Hintz, Carol, 166–67
Hintz, Raymond, 161, 164–65, 166, 167, 173
Hoberman, John, 203

Holmes, Oliver Wendell, 64
homosexuality, 82–83, 113–14
  conversion therapy and, 113–14
  testosterone and, 200
hormonal swings, menopause and,
  176
hormone-based rejuvenation drugs, 88
hormone-crime theory, 59
hormone replacement therapy, xiv
  blood clots and, 188–89
  breast cancer and, 188–89, 191
  cancer and, 187, 188–89, 191
  compounded hormones, 189–91
  estrogen and, 186–89, 191–92
  guidelines issued by North American
    Menopause Society, 191
  heart disease and, 187–88, 191
  history of, 183–86
  hot flashes and, 175–77, 191–92
  intrauterine devices for, 189
  patches for, 189
  pills for, 189
  progesterone and, 175–77, 186–89,
    191–92
  side effects of, xv
  stroke and, 188–89
  studies of, 187–89
  uterine cancer and, 186
hormone research, early, 10–12. See
    also specific researchers, conditions,
    and therapies
hormones, 14–33. See also specific
    hormones
  antibodies and, 154–57
  appetite and, 247–56
  brain stimulation and, 80–81
  chemical definition of, 7
  concept of, 36
  criteria for definition, 31
  discovery of, 6–7, 9–10
  disease and, 204 (see also specific
    conditions)
  hormonal defects, 247 (see also
    specific conditions)
  hormonal status, 115 (see also specific
    conditions)
  hormone-antibody bond, 154–57

hormone "type" or predominance,
    61–62
  immune cells and, 154–56
  impact on the brain, 66
  interconnectedness and, 258
  introduction of the term, 28–31
  killer, 54–71
  measurement of, xv, 153, 156
  prequel to, xiv
  reactions among, xv
  of sexual development, 89–101
  side effects of, xv
  Starling's introduction of the word
    hormone, 28–29
hormone tests, standardization of, 207
hormone therapy, 57, 125. See also
    hormone replacement therapy;
    specific hormones; specific therapies
    and conditions
  to accelerate menopause, 241
  age guidelines and, 242–43, 242n
  banning of human growth hormone
    therapy, 167
  blood sugar and, 160–61
  children and, 242–44
  contamination and, 160–73
  growth hormones and, 160
  height and, 126–47
  history of, 209–10
  immune response and, 153–54
  insulin and, 160
  marketing of, 201–3, 207, 208, 209
  side effects of, 242
  thyroid hormones and, 160
  for transgender men, 241–42, 244
  transgender treatment and, 231–34,
    241–44
  for transgender women, 242
Hoskins, Roy G., 57
hot flashes
  estrogen and, 175–77, 182, 184
  extracts from cow and sheep ovaries
    and, 184
  hormone replacement therapy and,
    175–77, 191–92
  research on, 178–80
  in whales, 180–81

HPV (human papilloma virus), 176
hugging, 229
Hulbert, Harold, 66, 68, 69, 70
human chorionic gonadotropin (hCG),
    99
Human Growth Foundation, 139
human growth hormone, 133–47
    banning of, 167
    extraction of, 142, 146–47, 171
    lawsuits and, 172
    synthetic, 167–68
    tracking of recipients, 171–72
Human Rights Watch, 117
hunger, 246–47
    hunger hormone, xiv, 247–56
    physiology of, 248–56
Hunter College, 152
hyperpituitarism, 43
hypogonadal axis, 205
hypophysenvorderlappenreaktion, 94
hypopits, 127
hypopituitarism, 43, 132, 145
hypopituitary dwarfs, 127
hypothalamic neurons, 182
hypothalamus, 8, 108, 181–82, 237
    growth and, 128
    leptin receptor in, 252–53
    as "mother-of-all-glands," 39
    obesity and, 246–56
    oxytocin and, 218, 221
    pituitary gland and, 39
hypothyroidism, 158–59
hysterectomies, 175–77, 191, 192–93

imaging techniques, xiii, 46, 133, 162,
    165, 238. *See also* brain imaging;
    specific techniques
immune cells, 9, 19, 154–56
immune response, hormone therapy
    and, 153–54
INAH3 (interstitial nucleus of the
    anterior hypothalamus), 237
"incidentalomas," 46–47
infants, 116. *See also* newborns
    crying, 227
    hypothyroidism and, 158–59
    sex of, 102–25

sex-switching surgery and, 111–12,
    116, 121
infertility, transgender treatment and,
    243
informed consent, 105–6
insulin, 57, 153
    animal source of, 136
    discovery of, xiii, xiv
    hormone therapy and, 160
    measurement of, 154
insulin antibodies, 154
InterACT, 117
interconnectedness, 258
"internal secretions," xii, 29, 64–65
International Olympic Committee,
    200
intersex children, 102–25, 236
Intersex Society of North America,
    121–22, 198
intersexuality, 102–25, *124*
    Johns Hopkins Hospital and, 235
    sex-switching surgery and, 111–12,
    116–17
interstitial cells, 74, 88
intrauterine devices, 189
in vitro fertilization, 89, 99
islets of Langerhans, 8

Jackson Laboratory, 249
James, Geraldine, 32–33
Jerome, Jerome K., 26
Job, Jean-Claude, 170
Johns Hopkins University/Hospital, 37,
    47, 90, 91, 97–99, 104
    Colapinto's articles on Money and,
    118
    hormone therapies pioneered at,
    112
    interdisciplinary approach at,
    112–13, 235–36
    intersexuality and, 235
    Johns Hopkins Gender Identity
    Clinic, 235–36, 244–45
    Office for Psychohormonal Research,
    113
    pituitary gland collection at, 144,
    144*n*

sex-switching surgery at, 116–18
transgender surgery and, 235–36
Jones, Duncan, 15
Jones, Georgeanna Seegar, *90*, 108–9, 110, 236
  CAH and, 112–13
  sex hormone research and, 89–101
Jones, Howard W., Jr., *90*
  CAH and, 112–13
  intersexuality and, 108–9, 110, 116
  sex hormone research and, 89–101
  transgender surgery and, 235–36, 244–45
Jordan-Young, Jordan, 111
Jorgensen, Christine, 232–35, *232*
Jorgensen, George, 232–33, *232*. *See also* Jorgensen, Christine
Jost, Alfred, 111
Joyce, James, 60
Joyce, Lucia, 60

Karkazis, Katrina, 118
Kennedy, John F., xvi
Khera, Mohit, 207
kidneys, 8, 10, 62
Kinsey, Alfred, 82*n*
Kinsey Institute, 223
Kipling, Rudyard, 26
Klopfer, Peter, 212–13, 219–20, 221
Koch, Fred, 197
Koch, Robert, 9, 30–31
Kraus, Karl, 83
kuru, 164

Lacks, Henrietta, 96
lactation, 8, 47, 218, 219
Laqueur, Ernst, 198
Latchmere Gardens, 31
Laurent, Bo, 123–25, *124*, 198, 236. *See also* Chase, Cheryl; Sullivan, Bonnie; Sullivan, Brian Arthur
Lay, Sarah, 170
Leibel, Rudy, 251, 254
Leopold, Nathan "Babe," 54–55, 61, 66–71
leptin, 3, 251–56
leptin deficiency, *250*, 251

leptin receptor defects, 252–53
Leydig cells, 74
Li, Choh Hao, 135–36
libido. *See* sex drive
Lichtenstern, Robert, 84
Lillie, Frank, 93, 110
Lind af Hageby, Lizzy, 23, 25–26
Lipocine, 207
Liquid Trust, 224
Lisser, Hans, 74, 226–27
Loeb, Richard "Dickie," 54–55, 61, 66–71, *67*
longevity, 208
love, 220–23
Lovelace, Linda, 114
"Low T." *See* low testosterone ("Low T")
low testosterone ("Low T"), 202–4
  Alzheimer's disease and, 209
  definition of, 206, 208
  heart disease and, 208–9
  screening for, 211
  testosterone therapy and, 205–6
"Low T" industry, 208–9
lutein, 128*n*

Macfadden, Bernarr, 76
MacGillicuddy, Adolf, 15
MacGillivray, Margaret, 167
mad cow disease, 164
Mahler, Fred, 145, 169
Maisel, Albert Q., 11
male climacteric, 202
male menopause, 195, 202, 203
maleness/manhood, 92–93, *199*
Malta, 117
Manson, Joann, 188–89, 190
Marie, Pierre, 135
Maryland Club, 98–99
maternal bonding, 217–23, *222*, 227
Mayo Clinic, 46
McElree's Wine of Cardui, 184, *185*
McGee, Lemuel Clyde, 197
medicine
  development of, 8–9
  status of, 74
Mencken, H. L., 63

menopause, xiv, 22
  acceleration of, 241
  hormone replacement therapy and,
    174–93
MenoPro, 192
menstruation, 60, 123
metabolic rate, 66, 69
metabolimeter, 66
metabolism, 8, 66, 148–49
microbiome, 253
micropenis, 103–4, 106, 115, 116, 118,
  120
Minkin, Mary Jane, 188–89
Mitchell, Roby, 209
Modlin, Irvin, 22
Money, John, 113, 129, 236
  Colapinto's articles on, 117–18
  criteria for treating children born
    with ambiguous genitalia, 114–15
  defintion of *gender role*, 115
monogamy, 114
Montagnier, Luc, 172
mood swings, 8
"moral molecule," 224–25
Morley, John, 202–3
Mortimer, Stanley, Jr., 38
Moses, David, 4–6
MRIs, 238
Müller, Peter, 111
Mullerian ducts, 111
multigland syndrome, 69
muscles, testosterone and, 204, 205

National Anti-Vivisection Society, 26
National Institute of Diabetes and
  Digestive and Kidney Diseases,
  169
National Institutes of Health, 144, 165,
  168–69, 171–72, 177–78
National Pituitary Agency, 139,
  144–45, 160, 165, 171
Nave, Gideon, 225–26
nerve theory, 78–81
nervous system, 7
neurokinin-B, 182–83
neuropathology, 47
neurosurgery, xiii, 38

newborns, 116
  hypothyroidism and, 158–59
  sex of, 102–25
  sex-switching surgery and, 111–12,
    116, 121
New York Endocrinological Society, 59
New York University, 227
North American Menopause Society,
  190, 191, 192
Novak, Emil, 99
nurturing, 220–23. *See also* maternal
  bonding

obesity
  hormones and, 1–13
  hypothalamus and, 246–56
Oliver, George, 12
Organon, 198
organotherapy, 62, 208
orgasms, 223
Osler, William, 9, 37
osteoporosis, 183
Ottoman Empire, eunuchs in, 58
ovaries, xiv–xv, 8, 10, 18, 30, 47, 57,
  81–82, 111
overeating, 246–56
oxytocin, 47, 214–29
  animal experimentation and, 217–23
  breastfeeding and, 223
  isolation of, 217–18
  maternal bonding and, 217–23, 227
  orgasms and, 223
  social skills and, 227–28
  synthetic, 215–16
  trust and, 223–25
  vaginal stimulation and, 220–21

Paley, William S., 38
pancreas, 8, 18, 20–21, 30, 62
panic attacks, 58
parathyrin, 60
parathyroid hormone (PTH), 60
parathyroids, 8, 60
parenting, 8
Parlow, Albert, 146–47, 171
Pastuszak, Alexander, 205
patents, 156

PATH (Partnership for the Accurate Testing of Hormone), 207
Patients' Bill of Rights, 105
Pavlov, Ivan, 19, 20, 21
Pedersen, Cort, 220–21, 228
pediatric endocrinologists, 165, 167. *See also specific researchers*
pedophilia, 114
penis. *See* micropenis
PEPI, 187
peristalsis, 19
Pettit, Michael, 77
Phall-O-Meter, *124*
pharmaceutical companies, 184–85, 190, 191. *See also specific companies*
  articles sponsored by, 203
  marketing and, 201–3, 207, 209
  research sponsored by, 203
  testosterone therapy lawsuits and, 211
Pharmacy Compounding Accreditation Board, 191
phrenology, 67
physiology, 8–9
pineal gland, 8, 69–70
Pitocin, 215–16
pituitary gland, 8, 18, 30, 39, 57, 108, 128*n*, *141*, 181, 257
  collection of pituitary glands, 134–47, 258
  Cushing's theory of, 39–40, 45–47, 94
  ergot and, 217–18
  function of, 47
  growth and, 128
  in human growth hormone, 134–47, 144*n*, 164–66, 170–71
  hypopituitarism and, 133
  hypothalamus and, 39
  as "mother gland," 39
  organotherapy and, 62
  oxytocin and, 217–18
  pregnancy and, 94–95, 98–99
  pregnancy tests and, 94
  three kinds of cells in, 45
  tranplantations of, 40

pituitary tumors, 42–44, 45*n*, 46–47, 69
  benign, 46
  symptomless, 46
  testosterone and, 204
placebo effect, 76, 85–86
placenta hormone theory, 96–98, 99
pneumoencephalograms, 133
polio vaccine, 163
"polyglandular" syndrome, 43–44
POMC deficiency (proopiomelanocortin deficiency), 247
pornography, 113, 114
posterior pituitary, 39, 41, 47, 217–18
posterior-lobe pituitary extract, 217, 218
Pound, Ezra, 60
pregnancy
  hormonal changes in, 186
  pituitary gland and, 94–95, 98–99
  posterior-lobe pituitary extract and, 217–18
pregnancy tests, xv, 93–96, 94*n*
prenatal hormones, gender identity and, 115–16
progesterone, 57, 185–86
  hormone replacement therapy and, 175–77, 186–89, 191–92
  hot flashes and, 175–77
  isolation of, xiv
  maternal bonding and, 220–21
  postmenopausal declines in, 22
progestin, 185
prolactin, 47
proles, 99
Prometrium, 189
Provera, 185, 188
pseudohomosexual behavior, 83
psychoanalysis, 233
psycho-endocrinology, 59, 65–66, 72, 113
psychopathology, 113
puberty, 115, 200, 253
puberty blockers, 242–43
publicity, placebo effect and, 76

quality of life, 208

Raben, Maurice, 137
radioimmunoassay (RIA), 155–59
Raiti, Salvatore, 140, 144, 171
Rance, Naomi, 181–83, 192
Reis, Elizabeth, 117
rejuvenation, 200
    testosterone shots for, 195, 208
    vasectomies for, 200
remedies, 62
    folk, 216–17
    herbal, 215
    quack, xiv, 30, 73–74, 258
reproductive endocrinology, 89–101.
        See also intersexuality; meno-
        pause; testosterone therapy
research. *See also* animal research;
        *specific researchers, institutions, and
        conditions*
    early, 10–12
    transparency in, 203
Riddle, Oscar, 129
risk-benefit analysis, xv
Roaring Twenties, xiv–xv, 56–57, 59,
        61, 73–74, 76
Rockefeller University, 250–51
Rodriguez, Joey, 160–63, 164, 166–67,
        173
Rodriguez, Ms., 161
roller tube machines, 96–98
Roosevelt, James, 37
Rothenberg, Ron, 208–9
Rothman, David, 129
Rothman, Sheila, 129
Royal College of London, 28
Royal Society, 20–21, 22–23, 31
Ruzicka, Leopold, 200–201

Sadler, William, 64
Safer, Joshua, 238, 242
salivary gland, 21–22, 24
Samuel, Larry, 169–70, 173
Sanger, Margaret, 63
satiety, 248–49
"satiety factor," 249
satiety hormone, 249–50
Schäfer, Edward, 29–30, 30n
Schäfer, James William Henry, 30n
Schartau, Leisa Katherina, 23, 25–26

Schering, 200–201
schizophrenia, 227–28
Schmidt, Peter, 87
Schneider, Bruce, 249–50
Schwartz, Al, 91
Schwartz, Lisa, 204
Scopes, John, 55
scrapie, 170–71
scrotum, 111
Searle, 185
Second International Congress on
        Eugenics, 63–64
secretin, 20, 22
secretions, 20–21, 29
    "internal," xii, 29, 64–65
    temperament and, 58
Seegar, G. Emory, 97. *See also* Jones,
        Georgeanna Seegar
self-help books, 76
seminal vesicles, 77–81
set point theory, 252
sex, 8
    assignment of, 102–25
    biology of sex differentiation, 92–94
    emotions and, 80–81
    of rearing, 115, 116
Sex and Internal Secretions (ed. Allen
        and Doisy), 92–94
Sex Club, 92–94, 110
sex drive, 70, 75, 78–81
    estrogen and, 207
    vs. sexual orientation, 113
    testosterone and, 200, 201, 204
    testosterone therapy and, 196, 206
sex gurus, 114
sex hormones, 89–101, 253. *See also
        specific hormones*
    business of, 88
    genitalia and, 108
    growth and, 128
    as growth hormones, 198
sex-switching surgery (for intersex
        infants), 107, 108, 111–12, 116,
        118–19, 124. *See also* transgender
        surgery
sexual anatomy, gender identity and,
        237–38. *See also* genitalia; *specific
        organs*

sexual behavior, 237
sexual orientation, 237
  vs. sex drive, 113
sexual prowess, testosterone and,
  194–213
*Shambles of Science: Extracts from the
  Diary of Two Students of Physiology*
  (Lind af Hageby and Schartau),
  25–26
Shapiro, Jake, 96
Sharpey, William, 30*n*
shortness, 126–30
Sing-Sing prison, 65–66
skepticism, 259
sleep, 8
sleep–wake cycles, 8
smell, sense of, 228
Snell, George, 249
Snizek, Karen, 246–47, 255–56
Snizek, Nate, 246–47, 252–53,
  255–56
Sobel, Edna, 131–32, 133, 134–35,
  137–38, 142, 156
social skills, oxytocin and, 227–28
Solitaire, Gil, 48, 144*n*
Solomon A. Berson Research Laboratory, 156–57
Spencer, Dennis, 38, 50
sphygmomanometers, 66
Stanford University, 161, 223
Starling, Ernest, 15, 18–25, 27–31, 33,
  36, 218
Starling, Gertrude, 18
Starling's Law, 19
starvation, 251–52
Steinach, Eugen, 72–73, 75–88, 200, 206
Steinach procedure, 84–88, 85, 200, 206
sterilization, compulsory, 64
Straus, Eugene, 152
stress, 8
stress hormone, 47
stria terminalis, 237
strokes
  hormone replacement therapy and,
    188–89
  testosterone therapy and, 211
subconscious desires, 59
Sullivan, Arthur, 102–3, 125

Sullivan, Bonnie, 111–12, 118–25. *See
  also* Chase, Cheryl; Laurent, Bo;
  Sullivan, Brian Arthur
Sullivan, Brian Arthur, 102–25
Sullivan, Cathleen, 102–12, 118–19,
  125
Sullivan, Mark, 105
Supreme Court, 64

temperament, secretions and, 58
testes, xiv–xv, 8, 10–11, 18, 29, 30, 47,
  57, 81–82, 112
  development of, 111
  injury to, 204
  organotherapy and, 62
  secretions from, 75
testicle-juice shots, for rejuvenation,
  75
testicles. *See* testes
testosterone, xiv–xv, 47, 112, 206
  aging and, 196–97
  athletes and, 200
  blood clots and, 211
  cognition and, 206
  deficiency in, 208
  estrogen and, 207
  exogenous, 205
  extraction of, 200
  fat and, 205
  genitalia and, 108
  gestation and, 115
  homosexuality and, 200
  injection of (*see* testosterone shots;
    testosterone therapy)
  isolation of, 87–88, 197–98
  levels of, 87, 196–97, 204–7
  libido and, 200, 201
  made from cholesterol, 200–201
  made from scratch, 199–200
  market for, 201–13
  mass-produced, 199–200
  menopause and, 176
  muscle mass and, 205
  muscle tone and, 204
  naming of, 198
  pituitary tumors and, 204
  production by Leydig (interstitial)
    cells, 74

testosterone (*continued*)
  for rejuvenation, 208
  sex drive and, 204
  sexual prowess and, 194–213
  testosterone deficiency (*see* low
    testosterone ["Low T"])
  transgender treatment and, 233
testosterone endopreneurs, 194–213
testosterone gels, xiv, 201–2, 205, 241
  advertisements for, 201–2
  off-label uses of, 204
  side effects of, 202
  skyrocketing sales of, 203
testosterone shots, 201–13
  for dwarfism, 132
  for rejuvenation, 195
testosterone therapy, 194–213. *See also*
    testosterone gels; testosterone shots
  animal experimentation and, 212–13
  benefits of, 205–6, 208
  energy and, 206
  heart attacks and, 211
  heart disease and, 205
  increase in blood cells and, 205
  lack of evidence for, 207–8
  lawsuits and, 211
  libido and, 196
  low testosterone and, 205–6
  marketing of, 201–3, 207, 208, 209
  off-label uses of, 204
  sex drive and, 206
  side effects of, 242
  strokes and, 211
  transgender men and, 241–42, 244
theory of evolution, 9
theory of hormonal "negative feed-
    back," 249
Thorne, Van Buren, 85
thymus, 30
thyroid, 8, 10, 18, 253
  defective, 12
  organotherapy and, 62
thyroid hormones, xiii, 3, 132–34, 253,
    258
  deficiency in, 148–49, 160
  hormone therapy and, 160
  metabolic rate and, 66

thyroid-stimulating hormone, 47
thyroid tablets, 148–49
Time magazine, 44–45
Todd, Lord, 217*n*
toxins, 253–54
transgender identity, 230–45
  suicide and, 245
Transgender Medicine Research Group
    at Boston University, 238
transgender men, hormone therapy
    and, 241–42
transgender surgery, 233–34
transgender treatment, 230–45
  hormone therapy and, 241–44
  infertility and, 243
transgender women, hormone therapy
    and, 242, 243–44
transitioning, 230–45
transmissible spongiform encephalopa-
    thies, or TSEs, 164
"transsexuals," 235
"transvestites," 235
trigeminal neuralgia, 38
trust, oxytocin and, 223–25
Trust Game, 223–25
Tufts University, 137, 142–43

United Nations, condemnation of geni-
    tal surgery on intersex infants, 117
University College London, 14–16, 24
University of California, Berkeley,
    142–43, 195
University of California, Los Angeles,
    146–47
University of Cambridge, 220, 248
University of Illinois, 150–51
University of Maryland, 144
University of North Carolina, 221, 228
University of Pennsylvania, 104
University of Zurich, 225
Upjohn, 185
U.S. Food and Drug Administration,
    165, 167, 186, 190, 203–4
uterine cancer, 175–76, 186, 192
uterus, 175
  contractions of, 217–19
  development of, 111

vaginal dryness, 192
vaginal stimulation, oxytocin and,
 220–21
vasectomies, 72–88
 for rejuvenation, 72–73, 75–76, 77,
  84, 85, 86–87, 200 (see also Steinach procedure)
 sex drive and, 72–73, 75–77, 83
vasopressin, 47
Vesey, William T., 29
vesicle–nerve theory, 78–79
vivisection. See animal research; animal
 vivisection
Voronoff, Serge, 74

Wahl, Chris, 49, 50, 51
warning labels, 190
Wasson, Sam, 179
weight. See also weight loss
 obesity, 246–56
 physiological studies of, 247–48
weight loss
 sustainable, 252
 weight loss surgery, 254
Weintraub, Arlene, 209
Wellcome Physiological Research
 Laboratories, 216
West Hudson Hospital, 102
whales, hot flashes in, 180–81
Whitney, John Hay, 37
Wilhelmi, Alfred, 143
willpower, xiv

Wilson, Alfred, 86–87
Wilson, Robert, 184–85
Wilson Foundation, 184–85
Woloshin, Steven, 204
Women's Health Initiative, 187–89
Woodward, Louis, 31–32
Wooldridge, Florence, 19
Wooldridge, Leonard, 19
Wymore, Mel, 230–31, 237–41, 239,
 243, 244, 245

X-rays, 46, 65–67, 69, 133
XY fetuses, 112

Yale University, 34–35, 38, 47, 219
 brain restoration project, 50–51
 Cushing Brain Tumor Registry,
  34–35, 38, 47, 48–53, 52
 Cushing Center, 34–36, 37, 51–53
 Grace–New Haven Hospital, 194–95
Yalow, Aaron, 151–52
Yalow, Rosalyn, 149–57, 158, 249–50
 death of, 158
 wins Albert Lasker Medical Research
  Award, 157
 wins Nobel Prize, 157
Yeats, William Butler, 73, 76
Young, Larry, 226–27, 228

Zak, Paul, 224–25
Zondek, Bernhard, 94–96, 98–99
zygotes, 93, 110